Constructing Futures

Constructing Futures
Industry leaders and futures thinking in construction

Paul Chan

Lecturer in Project Management, School of Mechanical,
Aerospace and Civil Engineering, University of Manchester, UK

Rachel Cooper

Professor of Design Management, Lancaster Institute
for the Contemporary Arts, University of Lancaster, UK

WILEY-BLACKWELL

A John Wiley & Sons, Ltd., Publication

Blackwell Publishing was acquired by John Wiley & Sons in February 2007. Blackwell's publishing programme has been merged with Wiley's global Scientific, Technical, and Medical business to form Wiley-Blackwell.

Registered office
John Wiley & Sons Ltd, The Atrium, Southern Gate, Chichester, West Sussex, PO19 8SQ, United Kingdom

Editorial office
9600 Garsington Road, Oxford, OX4 2DQ, United Kingdom
The Atrium, Southern Gate, Chichester, West Sussex, PO19 8SQ, UK
2121 State Avenue, Ames, Iowa 50014-8300, USA

For details of our global editorial offices, for customer services and for information about how to apply for permission to reuse the copyright material in this book please see our website at www.wiley.com/wiley-blackwell.

Library of Congress Cataloging-in-Publication Data
    Chan, Paul, 1976-
    Constructing futures : industry leaders and futures thinking in construction / Paul Chan, Rachel Cooper.
        p. cm.
    Includes bibliographical references and index.
    ISBN 978-1-4051-5797-1 (alk. paper)
    1. Construction industry–Great Britain.   2. Sustainable construction–Great Britain.
    3. Leadership–Great Britain.   I. Cooper, Rachel, 1953-   II. Title.
    HD9715.G72C43 2010
    624.068'4–dc22

                                                                    2010029192

A catalogue record for this book is available from the British Library.

This book is published in the following electronic formats: ePDF [9781444327847] and Wiley Online Library [9781444327830]

Set in 10/12pt Minion by Thomson Digital.
Printed and bound in Malaysia by Vivar Printing Sdn Bhd
1   2011

This book is dedicated to our loved ones and all those who have influenced our thinking.

Contents

About the authors   xi
Preface   xiii

Part 1   Tracing the past   1

1   Introducing foresight in construction: exploring the missing link of personalising futures thinking   3
Chapter summary   3
Setting the 'scene'   4
Construction foresight studies   7
    The 'future' agendas since the Second World War   7
    Contemporary foresight reports at the turn of the century   9
    Synthesis of critical trends and implications from the foresight reports   16
    Critical appraisal of foresight studies   17
A note on the methodology   22
The structure of the book   24
    Part 1: Tracing the past   24
    Part 2: Eliciting the future   24
    Part 3: Towards an afterthought   25

2   Influential people in the UK construction industry: what makes leadership in construction?   26
Chapter summary   26
Introduction   27
Leadership theories and application in construction   28
    Development of leadership theories   28
    Application of leadership theories in construction management research   30
    Looking forward but learning from the past   30
So what makes a 'leader' in construction?   31
    Critical antecedent: people   31
    Critical antecedent: place   34
    Critical antecedent: events   35
    To be or not to be? The age-old question of nature and nurture   37
Closing thoughts   41

Part 2   Eliciting the future   45

**3   Developing a sustainable future: theoretical and
practical insights into sustainable development**   47
  Chapter summary   47
  Introduction   49
  Connecting people, profits and planet: the rise of the
  sustainability agenda   50
    Interactions between people and places   51
    Role of political leaders and infrastructure development   58
    Industry response to the sustainable development agenda   68
    Role of education and research   70
    Summing up the thoughts of our leading figures   75
  Sustainability: definitions and perspectives   78
    Man-made capital: problems with an output-driven model   78
    Human capital: the rhetoric versus reality of investing in people   81
    Natural capital: consensus gained or paradise lost   86
    Social capital: building trust and sustainable communities   94
    The measurement problem: are efforts towards sustainable
    development doomed to fail?   100
  Closing thoughts   102

**4   Connecting up government, corporate and community stakeholders
in governing the future of the construction industry**   105
  Chapter summary   105
  Introduction   107
  Governance of the industry: seeking an institutionally coordinated
  response to meet the challenges of the future   108
    Think global act local   109
    The changing role of government: relinquishing control to the
    private sector   115
    Public–private interface   122
    People and managing relationships in construction   128
    Bringing interactions to the fore: exploring the intersections
    between government, corporate and community actors   140
  Shifting perspectives of governance   142
    Political governance: governance without government   143
    Corporate governance: the rise of corporate social responsibility   145
    Community governance: revisiting social capital   147
    The need for joined-up governance   150
  Governance in construction: the trends of privatisation and
  community engagement   155
    The relationship between government and construction   155
    Structure of the industry   161

Changing landscape of professionalism 165
Connecting political, corporate and community governance
in construction: the importance of human relations 172
Closing thoughts 173

Part 3   Towards an afterthought 177

5   The last word: synthesising lessons learnt from the journeys... ... 179
Chapter summary 179
Introduction 180
Recap on previous chapters 181
Leadership as an emergent process: moving away from
individualistic notions of leadership in construction 181
The missing social link of sustainable development 182
The aspirations of joined-up governance 182
Key conclusions 183
Futures thinking as emergent thinking 183
Disrupting boundaries: the age of hybrids 184
Tensions, ambiguities and paradoxes 185
Knowledge gaps to frame the research and practice agenda
of the future 187
Epilogue 187

Appendix: Brief biographies of influential figures interviewed ... 189
Alan Ritchie 189
Bob White 189
Chris Blythe 190
Chris Luebkeman 190
George Ferguson 190
Guy Hazelhurst 190
Jon Rouse 191
Kenneth Yeang 191
Kevin McCloud 191
Nick Raynsford 192
Sandi Rhys Jones 192
Sir Michael Latham 192
Stef Stefanou 193
Tom Bloxham 193
Wayne Hemingway 194

Notes and references 195
Index 219

About the authors

Paul Chan is Lecturer in Project Management at the School of Mechanical, Aerospace and Civil Engineering (MACE) at the University of Manchester. His research interests span the areas of human resource management and development in project-based environments, dynamics of migrant employment in the construction industry, interorganisational relations, organisational learning and performance management. He has undertaken funded research projects on organisational knowledge and productivity, skills capacity issues, and design decision-making in infrastructure development. He is also enthused by the broad area of sustainable development and takes a keen interest in the personal development planning of students. Paul is an incorporated member of the Chartered Institute of Building, a committee member of the Association of Researchers in Construction Management (ARCOM), and a joint coordinator of the CIB Task Group 78 on 'Informality and emergence in construction'. He is also a Fellow of the Higher Education Academy, and a member of the European Institute of Construction Labour Research (CLR) Great Britain, and CIB Task Group 59 on People in Construction. He has authored around 50 peer-reviewed journal and conference articles to date.

Rachel Cooper is Professor of Design Management at the University of Lancaster, where she is Director of the Lancaster Institute for the Contemporary Arts and also ImaginationLancaster (a centre for research into products, places and systems for the future). Her research interests cover design management; design policy; new product development; design in the built environment; design against crime and socially responsible design. She has led on numerous UK government-funded research projects, spanning a range of design-related issues, including requirements capture, future scenarios for distributed design teams, design against crime, use of design in government departments, and process thinking and management. She also led on the £3m Engineering and Physical Sciences Research Council (EPSRC) funded 'Vivacity 2020' research programme investigating sustainable urban design for the 24 hour city. She is currently co-investigator of Urban Futures, a five-year EPSRC-funded research project (see www.urban-futures.org). Rachel has authored several books in the field including *The Design Agenda* (1995), *The Design Experience* (2003) and edited *Designing Sustainable Cities* (2009). She is currently commissioning editor for an Ashgate series on *Socially Responsible Design*. She was commissioned in May 2007 by the Foresight programme on Mental Wellbeing and Mental Capital to write the *Scientific Review on Mental Wellbeing and the Built Environment*. Professor Cooper is President of the European Academy of Design, and Editor of *The Design Journal*. She has also undertaken several advisory roles to national and international universities, government and non-governmental organisations.

Preface

This book is as much about constructing relationships as it is about constructing an understanding of 'futures thinking' in the construction industry. The genesis happened in 2004 when a new working relationship was constructed between the two of us – one the Professor and the other a new researcher. During a regular research meeting early that year, the Professor suggested that it would be a good idea for us to talk to the good and the great of construction to find out what cutting-edge issues were confronting the industry. This was to be the defining moment that marked the start of the journey that led to the production of this book. Our initial desire was to get closer to the leaders of industry to find out about the research agendas that mattered in practice. After all, being a non-cognate, the Professor had benefited herself in the mid-1990s from constructing relationships with senior people in a number of UK construction companies to gain a better insight into the world of construction. So, we thought that it was timely then to get a renewed perspective of the critical issues that industry leaders faced so as to help inform our research directions. Thus, we proceeded to write to a wide range of senior figures to see if they were interested in participating in our project.

By the middle of 2005, we constructed relationships with 15 influential figures in the UK construction industry. What started out as a few interesting conversations led us to take a number of fruitful detours. Increasingly, we became fascinated by the life stories that these influential people had to share with us, especially in terms of how their personal life experiences intimately merged into the career paths that they eventually undertook. We were also privileged to obtain their firsthand accounts of how the UK construction industry had developed from the past, the issues that were confronting the industry at the present and the challenges to be addressed for the future. The personal stories recounted by our influential participants provided rich, practical insights into what motivated them to continue working in the construction industry. Often, it is this underlying passion for construction that sustained their sense of optimism for the future. The issues discussed were wide-ranging and eclectic, and, at times, the messiness of the data created difficulties for our sense-making process. Nevertheless, it was the identification of these issues that drove us to search deeper into a variety of literature sources, from policy perspectives to foresight studies, to theoretical perspectives of leadership, sustainable development and governance, resulting in the final compilation presented here.

Two things really registered with us as we progressed through this 'project'. First, we were struck by how the legacies of the past, the lessons from the present and the thoughts about the future of the construction industry were so interwoven in the tales told by our willing participants. Second, we got great comfort from the conversations with our influential figures as there was much confluence between what they had to say and the theoretical literature that we have subsequently reviewed. This was, therefore, instrumental in the way we have designed the structure of this book. To help the reader navigate through our sense-making process, we have organised this book into three main parts. In Part 1, we look to the past by reviewing a series of foresight studies of the construction industry (Chapter 1), and re-present stories of our interviewees' lives to explain the development of leadership in the context of the construction industry (Chapter 2). In Part 2, we look to the present by discussing two fundamental issues that emerged in our analysis of the interview data. These two issues relate to the various perspectives of sustainable development (Chapter 3) and the governance of the construction industry (Chapter 4). In Part 3, we conclude with an afterthought for the future, highlighting key lessons learnt by embarking on this journey and putting forward a series of research questions derived from what we consider to be a scholarly reflection of 'futures thinking' in construction presented in this book.

Within each of Chapters 2, 3 and 4, we have juxtaposed the views of our influential figures with a review of the salient points found in the relevant and authoritative sources of theoretical literature, both in the mainstream literature and the field of construction management. Whereas we do not profess to treat the diverse range of literature bases exhaustively, it is hoped that the book will allow the reader to benefit from learning practical insights from our participants while gaining a rapid understanding of the key debates of the theoretical subject under scrutiny. By placing our analysis of the interviews adjacent to the theoretical review, it is also our intention to enable the reader to make comparisons so that overlaps and gaps between theory and practice can be ascertained.

It is hoped that this book will contribute to a number of areas within the broad field of construction management. First, the 'project' has permitted us to construct a deep reflection of the life histories of our 15 participants. Life history methodology is rarely deployed in the field of construction management, and, therefore, we hope that our analysis provides knowledge of not only what motivates our senior leaders in construction, but also how these value systems drive the way they think about the future. Specifically, we have identified how 'futures thinking' in construction is really emergent, shaped in part by the personal passions of leaders in construction. Thus, this challenges the orthodox view that 'futures thinking' in construction, in the form of scenario planning, must be strategic and objectivised. Second, and following on from the critical finding of emergence, our analysis has revealed the internal struggles that our influential figures go through when making sense of the contemporary agenda of sustainable development. This provides a rich picture of the dynamic ways in which our senior figures grapple with the grand challenges confronting the industry of the day. Such personal journeys are rarely reported in conventional research in the field. Indeed, the analysis has sought to explain the critical issues that help shape decision-making when talking about the creation of a sustainable future. The trade-offs are articulated in the form of tensions, ambiguities and paradoxes that our interviewees constantly seek to resolve. Leading on from this is the third point of our contribution. Much of the discussion on emergence implies the impotence of prescriptive methods, often espoused in the construction management literature. Indeed, boundaries as perceived by our influential figures tend to be fuzzy, as our practitioners get on with experimenting with new methods of working and new ways of organising relationships and governance structures in moving forward to the future. Such fuzziness does not mean the perspectives of our influential figures are any less valid than the neat categories formulated by theoretical endeavours. Rather, there is much scope in seeking cooperation between the academy and industry as we make sense of the future for the construction industry. As the title suggests, the book is about *constructing* the future of the industry as opposed to maintaining a *constructed* view of the future.

Finally, we would like to acknowledge a number of key people in helping us with the journey of *Constructing futures*. First, our gratitude goes to Madeleine Metcalfe and her editorial team at Wiley-Blackwell for their patience and encouragement over the years it took to produce the final manuscript. We would also like to acknowledge Professor Cary Cooper, whose work in the 1980s on *The change makers* inspired our endeavours in the first instance. We also recognise the financial support provided by the Engineering and Physical Sciences Research Council (EPSRC) in the UK (Grant references GR/R64575/01 and EP/D505631/1), which facilitated some of the early fieldwork necessary for this book project. We also like to extend our appreciation to Paula Richardson who provided immense assistance in the transcribing of the interviews, Roger Whitham for his efforts in producing the illustrations, and to Dr. Louise Bissett and Dr. Michael Pritchard for the conversations that helped in our thinking about the conclusions for this book. The ideas presented in this book have also been shaped through discussions with a wide range of colleagues, whom we have interacted with during various conferences. These included the CIB W55/W65 symposium in Rome 2006 and research seminars at Northumbria University, Chalmers University and the National University of Singapore, which helped in the development of Chapter 2; the 21st ARCOM conference in London that helped validate ideas put forward in Chapter 1, and ARCOM research workshop on 'sustainability' in Plymouth, which provided feedback for theoretical aspects reviewed in Chapter 3. Last, but not the very least, without our 15 influential figures and their willingness to offer up their time to share their stories with us, this book would never have come to fruition. We are, therefore, indebted to all of them. However, all errors and omissions remain our sole responsibility.

Paul Chan
University of Manchester

Professor Rachel Cooper
Lancaster University
February 2010

Part 1

Tracing the past

Chapter 1

Introducing foresight in construction: exploring the missing link of personalising futures thinking

'Man can believe the impossible, but can never believe the improbable.'

Oscar Wilde, 1854–1900

Chapter summary

This chapter reviews previous endeavours undertaken to shape the critical agendas and forecast future trends for the construction industry. We draw on material from various post-war inquiries and a number of foresight reports, primarily from the USA, UK, Australia and Sweden, to synthesise some of the emerging future trends and explore their implications. Our reflection suggests that future scenarios have tended to portray an optimistic future for the sector, as stakeholders grapple with the uncertainties and tensions arising from socio-technical change. Consequently, these scenarios serve to present alternative, somewhat ideal positions to which all those working in the industry can aspire. Much work in future scenario planning has relied on the astute few to undertake blue-sky thinking, usually in committees, and so 'futures thinking' presents only a partial view of what tomorrow's society might look like and how this impacts on the future workings of the industry. There is often a disconnection, in that efforts to predict the future tend to be detached away from those who are operating at the coalface of operational realities in construction. This book project is driven by our desire to go some way in 'personalising' futures thinking. We do so by tracing the journeys and personal motivations of those who tend to get involved in shaping the future agenda for the industry. This chapter sets the scene for the book by justifying the need to examine how leading figures of the UK construction industry have thought, and are thinking, about the future of the industry. In so doing, we present 'futures thinking' as an emergent process, rather than the objectivised positions often depicted in the reporting of future scenarios found in several foresight studies reviewed here.

Constructing Futures: Industry Leaders and Futures Thinking in Construction   Paul Chan and Rachel Cooper

The key issues discussed in this chapter are as follows:

- There are very few, if subtle, differences between agendas of the past and the future. Two recurrent themes appear to emerge in the review. First, efforts to shape the industry's agenda of the day are influenced by social change characterised by the transformation of power relations in the industry. Second, the focus on performance has often been critical in framing of the 'futures' agenda of the industry.
- Foresight studies undertaken about the construction industry often maintain a sanguine view on the role technology plays in developing the construction industry, and often downplay social considerations.
- 'Futures thinking' in the construction industry is, in many respects, an extension of thinking about present-day issues. This raises questions as to whether foresight reports in construction genuinely engender a strategic long-term view, both in terms of policy formulation and implementation.
- It is unclear how legacies of the past play a part in shaping our understanding of the future. There are certainly convergences between past agendas and the identification of future challenges. To connect the past, present and future, there is a need to delve deeply into the motivations of people who have an influence on the development of the construction industry. This forms the main thrust of this book.

Setting the 'scene'

'Oracles, futurists, visionaries [...] divine the shape of things to come before anybody else. And we all avidly await their predictions.'[1] Human societies have always been fascinated by the apparent ability of some to foretell what is going to happen in the future. From the prophecies of Nostradamus, to the daily reading of personal horoscopes, to the monitoring of weather forecasts, we have always been interested in what future trends may lie ahead so that we can better prepare ourselves in the present. As trend guru Faith Popcorn asks on her website www.faithpopcorn.com: 'If you knew everything about tomorrow, what would you do differently today?'. Indeed, the ability to accurately predict future trends can be financially rewarding for those trendsetters whose ideas endure over time. This is arguably the case for Faith Popcorn and the strategic trend-based consultancy BrainReserve founded in 1974, who claimed to have identified the 'caffeinated hegemony of Starbucks to the cultural squeeze on Wal-Mart to the explosion that is MySpace'. The prospect of owning a crystal-ball that can allow you to gaze into the future seems rather enticing, especially during the turbulent times of the present day. In fact, there is no better time to start strategising on what to do to confront the challenges of a financial crisis that has been beset upon us, as troubling times offer immense opportunities to steer into new, unchartered terrain. Despite the gloomy headlines surrounding the economic recession, the resilience of human societies means that there is an ironic sense of optimism and tenacity as questions are raised regarding how and when economic recovery will take place in the near future.

Increasingly, the art of forecasting is also becoming less mystical as more robust and rigorous techniques are being introduced. Faith Popcorn, for instance, bases her trends (see Box 1.1 below) on getting as close to the marketplace as possible; she does this by

Box 1.1. Having Faith in Future Trends – the trends spotted by Faith Popcorn (see also www. faithpopcorn.com)

Faith Popcorn founded BrainReserve, a strategic trend-based marketing consultancy, in New York City in 1974. Their work basically revolves around helping businesses identify how future trends might help shape the development of new products and services. Notwithstanding criticisms that question the scientific validity of trendspotting, what Faith Popcorn is doing with BrainReserve is effectively acting as a commentator of how societal trends are shaping over time. Below is a summary of a number of interesting trends that have been identified over three decades, and still endure in today's marketplace:

* *99 Lives*: where the pressures of time and the constant demand for fast delivery force people to multi-task;
* *Anchoring*: where there is a desire to hook on to something stable and secure in a fast-moving, ever-changing world;
* *Atmosfear*: where there is greater fear of environmental concerns;
* *Being alive*: where there is greater consciousness on the need to live healthy lifestyles;
* *Cashing out*: where there is greater awareness that career satisfaction is not the only goal in life, and that personal life satisfaction matters most;
* *Clanning*: where there is a greater need to feel a sense of belonging;
* *Cocooning*: where there is a greater desire to avoid the realities of the outside world (often credited for the growth in home-working and home-shopping);
* *Down-ageing*: where there is a greater desire to return to the experiences of being a child;
* *Ergonomics*: where there is a greater movement towards mass customisation;
* *Eveolution*: where there is increasing recognition of the value and power of women as consumers;
* *Fantasy adventure*: where there is greater need to experiment with new things or go through new experiences;
* *Futuretense*: where the pressures of today cripples the human ability to cope in today's world, let alone think about the world tomorrow;
* *Icon toppling*: where there is a growing movement to question established forms of authority;
* *Pleasure revenge*: where consumers increasingly revel in a lifestyle of excess;
* *Small indulgences*: where there is a growing market for affordable luxuries;
* *SOS (Save our society)*: where there is greater awareness for ethical concerns;
* *Vigilante consumer*: where the power shifts to the discerning consumer of the future.

interviewing thousands of consumers and tracking popular culture through the analysis of hundreds of media. In her own words, she explains the process: 'We watch. We listen. We intuit. We connect the dots. We continually ask ourselves, why am I seeing this? What is this connected to? What may this effect?'. This book focuses on the growing interest over the last decade in the development of future scenarios that are specifically generated for the construction industry. Numerous foresight studies have been produced. In general, these are often written in collaboration with academic and professional experts, together with client groups represented mainly by government stakeholders, to map out future trends and its implications across political, economic, social, technological, legislative and environmental dimensions. The framing of future scenarios that could impact on the construction industry is in many respects similar to the approach adopted by Faith Popcorn, i.e. undertaken by engaging with the marketplace of the construction industry.

In this book, we are less concerned about reiterating future trends and their implications. The reader is instead referred to the many foresight studies that exist, most of which, as we shall see in this chapter, paint a pretty consistent picture of future trends for the industry. Besides, ideas about the future can often be filled with so much optimism that there is a high probability that these do not translate to reality. Instead, the focus, as suggested in the title, is on how alternatives about the future are being constructed through the eyes of those who tend to get involved in the production of these foresight studies. As the review below will show, many foresight studies are created through the efforts of committees made up of a mixture of professionals working in the industry and client representation usually in the form of government officials. Yet the value systems of those involved in these committees are rarely articulated; 'futures thinking' has the tendency to be objectivised. This book aims to personalise 'futures thinking' by providing a deeper, more critical analysis of how thinking about the future is shaped through the eyes of a selection of influential people engaged within the UK construction industry. In so doing, it is hoped that this book will empower the reader to think about constructing their own future in the industry as we trace the personal journeys of some of the influential leaders in UK construction.

This introductory chapter is organised into three main sections. First, a salient review of key institutional reports into future scenarios for the construction industry will be presented. We begin this review by revisiting the 'futures' agenda of the past, as we reiterate the summary of many reports written about the UK construction industry since the Second World War.[2] Second, we contrast the past with a review of more contemporary foresight reports written at the turn of the twenty-first century. The intention is to illustrate just how similar and relevant some of the agendas framed previously are to the critical concerns raised in the present about the future. Such convergence is not surprising, as many of the reports are written by the established institutions of the industry. We argue that what is needed is a deeper exposition of the value systems of those contributing to the writing of these reports, in order to develop a more comprehensive understanding of how and what drives futures thinking in the construction industry. Third, and following from the review, the methodology supporting the data collection and analysis that informs the writing of this book is discussed, before a brief synopsis of later chapters is described.

Construction foresight studies

Over the last decade, there has been a growth in foresight studies across different countries and industries. The basic assumption governing these studies is that the 'future will be different from anything we have seen before' and that foresight studies represent an attempt 'to gauge how the longer-term future may manifest itself'.[3] The impetus, therefore, for understanding future scenarios is to better prepare the general population for uncertainties and challenges that lie ahead. As the chapter unfolds, it would appear that these foresight studies tended to be conducted in a top-down fashion, often based on select committees led by key representatives from institutional organisations in government and industry. In this section, a review of key foresight reports of the past and present is outlined, with a view to appraising some of the future trends and implications that have been identified as being pertinent to the construction industry. A critique is also provided on the usefulness of such foresight studies as objectivised futures, as we make the case for a more thorough appreciation of the emergent nature of futures thinking.

The 'future' agendas since the Second World War

Before one can delve into a commentary about the future, it is always useful to revisit the legacy of the past, as history does teach us a lot if we bother to find the lessons there. Since the Second World War, several governmental and institutional reports have been written about the UK construction industry. For a fuller discussion about these reports, you are encouraged to read Murray and Langford.[4] In this section, we have merely reiterated a summary of some of the recurrent themes established through their analysis of the reports of the past (Table 1.1).

Two fundamental issues consistently re-occur in these past reports. The first issue revolves around the configuration of relationships in the construction industry. Typically, authors of these inquiries have expended much energy in explaining various arrangements of the relationships between key stakeholders of the client, designer, contractor and supplier, and how the different ways of organising these stakeholders can bring about advantages and disadvantages to the design and construction process. The discussion also tended to revolve around contractual issues, and it is evident that there is a change in focus towards more relational aspects when discussing how relationships between stakeholders are being constituted. So, terms such as partnering, integration and supply chain management have gained a certain degree of familiarity in the discourse of today's construction both in the UK and globally. Furthermore, when examining the changing emphases in the political and institutional debates outlined in these reports, one can trace the shift of power away from the professionals working in the industry to the clients and users of the built environment, due in part to the rise of consumerism and consumer sovereignty.

A second recurring theme is the existence of a strong performance improvement agenda since the post-war period. So, the desire to raise the game in the industry is not a new phenomenon. However, specific emphases of the performance agenda evolve over time. Various crisis moments help shape the focus at a particular point in time, and perhaps explain why reports into the affairs of the industry have been commissioned in the first place. Admittedly, we can see, for instance, how the drive to boost rebuilding efforts at the

Table 1.1 Key reports on construction (adapted from: Murray and Langford, 2003: 5)

Report	Year	Title	Driver
Simon	1944	*Placing and management of building contracts*	Focuses on contracts and the need for less bureaucratic tendering in competitive tendering, particularly in government contracts
Phillips	1950	*Report on building*	Focuses on coordination and public clients seeking better performance through improvements in labour productivity and the management of the construction process.
Emmerson	1962	*Survey of problems before the construction industry*	Focuses on integration of the design and the construction process; giving rise to the popularisation of design and build
Banwell	1964	*Placing and management of contracts for building and civil engineering work*	Focuses on management of the building process and constructors' need for Government regulation of public contracts through negotiation
Tavistock Institute	1965 and	*Communications in the building industry*	Focuses on the systemic conflict that characterises the construction industry and
	1966	*Interdependence and uncertainty*	created a greater emphasis on the sociological perspective of construction
Wood	1975	*The public client and the construction industries*	Focuses on placing of public contracts via package deals (more negotiated work)
Latham	1994	*Constructing the team*	Focuses on relationships between the parties to the construction process
Technology Foresight	1995	*Progress through partnership*	Focuses on political, social and technical alignment of a changed agenda set by government
Egan	1998	*Rethinking construction*	Focuses on performance and productivity of the industry
Fairclough	2002	*Rethinking construction innovation and research*	Focuses on the industry's lacklustre approach to invest in research and development and the need for closer collaboration between industry and academia

end of the Second World War has helped frame the Simon report of 1944 into finding efficient ways of placing contracts and managing construction demand; and how the deterioration of industrial relations in the 1960s contributed to the sociological studies undertaken by the Tavistock Institute; and how the recession of the 1990s has forced a rethink about the adversarial working relations of the industry in the Latham report of 1994; and how the desire to adopt the performance agenda of the manufacturing industry drove the efforts of the Egan agenda in the late 1990s. In any case, much of the performance agenda also focused largely on how the construction industry copes with fluctuations in the economy, and so there has been, and still is, a great deal of interest since the post-war era of seeking the stabilisation of economic cycles for the benefit of the UK construction industry.

It is unsurprising to note that governments do take an interest, and play an intimate part, in their interactions with the construction industry. The industry does underpin many of the other economic sectors through the critical provision of the built environment by which other industrial sectors depend on to function effectively. Globally, the government is also a major client of the industry, often accounting for nearly half of construction output. This possibly explains why reports are constantly being commissioned to capture the state of affairs, as well as chart out future challenges confronting the sector. In the next subsection, we review more contemporary reports into the future scenarios that can impact on the construction industry.

Contemporary foresight reports at the turn of the century

Eight sets of foresight studies specific to the construction industry have been analysed for the purpose of this chapter. These were selected from the UK (three), USA (three), Sweden (one) and Australia (one), so as to maintain, as far as possible, an international view of the future. The organisations driving the publication of the selected reports and the details are summarised in Table 1.2.

A brief background to the remit of the seven organisations and their key recommendations on future trends is outlined here.

The Construction Research and Innovation Strategy Panel, UK (CRISP)

The Construction Research and Innovation Strategy Panel was originally formed in July 1995 as a joint industry and government panel to identify and develop priorities for research funders and help set the agenda for construction research and innovation in the UK. Renamed as nCRISP in 2002, following the Fairclough[5] report *Rethinking construction innovation and research*, nCRISP maintained the remit of prioritising and promoting research and innovation that would sustain a first class construction industry and enhance the value of its contribution to the quality of the built environment and the wealth and well being of society. The Panel had a widely based council that met twice a year, whose membership included construction clients, major industry bodies, government departments and agencies with a significant interest in construction and the built environment and the research community. In 2005, nCRISP was subsumed under the National Platform for the Built Environment (www.nationalplatform.org.uk), an industry-led group focused on promoting strategic research to industry and its wider stakeholders.

Table 1.2 Background information to the selected future reports

Organisation	Title	Author(s)/(Year)	Country
Construction Research and Innovation Strategy Panel (nCRISP)	*Building future scenarios*	Edkins (2000)	UK
	Constructing the future Nanotechnology and implications for products and processes	Broyd (2001) Gann (2003)	
Construction Industry Research and Information Association (CIRIA)	*UK construction 2010 – future trends and issues briefing paper*	Simmonds and Clark (1999)	UK
Commission for Architecture and the Built Environment (CABE)	*The professionals' choice – the future of the built environment professionals*	Royal Institute of British Architects (2003)	UK
Construction Industry Institute (CII)	*Vision 2020*	CII Strategic Planning Committee (1999)	USA
Civil Engineering Research Foundation (CERF)	*The future of the design and construction industry (projection to 2015)*	Building Futures Council (2000)	USA
American Society of Civil Engineers (ASCE)	*The vision for civil engineering in 2025*	ASCE (2007, 2009)	USA
	Achieving the vision for civil engineering in 2025: a roadmap for the profession		
Chalmers University	*Vision 2020*	Flanagan, Jewell, Larsson and Sfeir (2000)	Sweden
The Australian Cooperative Research Centre for Construction Innovation (CRC)	*Construction 2020: a vision for Australia's property and construction industry*	Hampson and Brandon (2004)	Australia

Nonetheless, the priorities included a wide range of issues, many of which are still relevant today. These encompassed customer needs, sustainable construction, design, technologies and components, process and performance, information and communication technologies (ICT), housing and construction research base, regulatory and financial framework and motivation and communication. From these priorities, it can be seen that nCRISP had a large focus on technology. In fact, as part of identifying future trends for the construction industry, the Panel embarked on the foresight programme in the late 1990s, which had as its aim the desire to increase the UK's exploitation of science. The programme ascertained either potential opportunities for the economy/society from new science and technology or how future science and technology could address societal challenges ahead. Three key reports were produced as a result, which included *Building future scenarios*[6] that led on to *Constructing the future*[7] and latterly a review of *Nano-technology and implications for products and processes.*[8] Based on the Social, Technological, Economic, Environmental and Political (STEEP) framework, as well as other wider foresight studies such as *The long boom*,[9] future scenarios[10] were developed by the researchers involved. The key emerging trends identified were as follows:

- Social: ageing population, rise of the urban population, restructuring the notions of the 'family', 'home' and 'work', and rise of individualism;
- Technological: use of ICT in facilitating a knowledge culture, use of biotechnology in materials, food and medicine, growth of nanotechnology, and alternative energy sources;
- Economic: shift towards the service industry, greater utilisation of human skills and technology, and consideration of the location of firms;
- Environmental: climate change, regional sea defences and water storage and supply, and levels of governance (i.e. local, regional and national) and its impacts on the environment;
- Political: layers of governance (see Environmental above), and the innovative use of the public purse;
- Specific to the built environment: globalisation and increased competition, provision for housing in terms of design, construction and use, implications of increased use of ICT in the workplace, development and use of sustainable materials, safe construction and refurbishment and reuse of buildings.

Establishment of these trends led to a number of recommendations, including:

- The promotion of 'smart' buildings and infrastructure;
- Improvement of health and safety;
- Enable supply chain integration;
- Invest in people;
- Improvement of existing built facilities;
- The need to exploit global competitiveness;
- The need to embrace sustainability;
- The need to increase returns on investment, and;
- The need for forward planning.

The Construction Industry Research and Information Association, UK (CIRIA)

The Construction Industry Research and Information Association is an independent research and innovation organisation in the UK with three main research foci: building and facilities, transport and water facilities. Their key concerns include technical issues, legislation and regulation, training, management and economics. Complementing the efforts of nCRISP, a research team at CIRIA embarked on a project aimed at eliciting future trends from industrial practitioners. This fulfilled part of the agenda of 'Adopting foresight in construction'. In 1999, Simmonds and Clark[11] reported on findings derived from interviewing more than 140 participants across eight companies (undisclosed and no mention was made regarding research methodology) and they concluded with the following implications for the construction industry:

- Increasing customer-centric focus;
- Types/use of buildings and shorter building life cycles;
- Rising importance of housebuilding and infrastructure;
- Increasing globalisation and international trade and competitiveness for the industry;
- Changes in planning and development in terms of restrictions on greenfield sites and rise of the self-build sector;
- Changes in construction processes with a growth in standardisation and prefabrication;
- Growth in skills and competence development; and
- Increasing importance of sustainable materials and use of land.

The Commission for Architecture and the Built Environment, UK (CABE)

The Commission for Architecture and the Built Environment is an executive non-departmental public body based in the UK with ongoing foci on educational and healthcare facilities, residential homes and a strong design remit. The Commission is primarily involved with engaging with young people where design of the built environment is concerned, the housing and regeneration agenda, the design and use of public space and environment, as well as skills and planning for the future of the industry. As part of initiating the debate about the future, a book entitled *The professionals' choice: the future of the built environment professions*[12] was published in 2003. This book contained several scenarios that were each taken up by a leading academic expert to 'imagine forward and wrote backwards'. These scenarios included:

- Regulatory: increasing self-regulation of the professions, professionals providing more 'consultancy' in risk management, flexible working;
- Economic: towards a service industry with user-centric focus, increasing agenda for environmental sustainability, growth in leasing rather than owning, need to rethink skills to meet ever-changing business models;
- Technological: decline in construction undergraduates, increase mass customisation and diminishing role of the professions, growth of alliances and supply chain integration;
- Social: increase personal autonomy and decline in traditional education in terms of career paths/choices, increasing need to be culturally sustainable, shifting definition of work and impact on personal lifestyles, rising importance of environmental and sustainability issues;
- Managerial: integration of construction professions, shift towards softer 'creative' skills and move away from hard engineering and management, construction becoming a

more stable sector due to shift towards offering the whole package of building and servicing.

The Construction Industry Institute, USA (CII)

The Construction Industry Institute based at the leading University of Texas at Austin, USA is a network of more than 90 organisations representing clients, contractors and suppliers in both the public and private sectors. Its main remit is to engage with these industrial partners to deliver business effectiveness and improvement of capital facilities over its life cycle through dealing with such matters as safety, quality, schedules, cost, security, reliability, operability and global competitiveness. To guide its long-range planning process, the CII's Strategic Planning Committee began a series of meetings and consultations in November 1997 to develop blue-sky thinking in a number of areas. These culminated in the production of a report entitled *Vision 2020* in 1999,[13] which is summarised as follows:

- Globalisation: intertwining of national and regional economies, rise in international procurement, increasing geographic dispersion of integrated teams, skills needed to align different cultures and interests, upgrading of technical competence in developing countries;
- Technology: increase use of ICTs blurring the lines between project phases, sustainable materials, automation on-site;
- Business relationships: increase in project alliances, more comprehensive project management skills, reshaping of business entities, changing stakeholders' roles, growing importance of risk management;
- Characteristics of projects: increasing focus on renovation and renewal, rising importance of flexibility, operations, maintenance and decommissioning becoming more crucial at the front-end, increase project complexity;
- Planning, design and construction practice: increased use of prefabrication and standardisation, enhancement in resource planning coupled with increased automation, rising importance of intelligent handheld systems, need for real-time performance measurement;
- Workforce: minimised use of craft labour and increase in capital substitution, growing need to consider work–life balance, importance of recruitment and retention of high-quality engineering graduates, increased industry–academia collaboration.

The Civil Engineering Research Foundation, USA (CERF)

The Civil Engineering Research Foundation forms part of the American Society of Civil Engineers (ASCE). Based in Washington D.C. USA, their main remit is to act as the engine for dissemination and application of research and innovation in the industry. The chief priorities of the CERF lie within the areas of productivity, performance and sustainability within design and construction through collaboration and innovation. The CERF also holds the directorship of the Building Futures Council (BFC), an organisation aimed at promoting future-oriented thinking across the American construction industry. In 2000, CERF produced a foresight report entitled *The future of the design and construction industry (projection to 2015),*[14] which identified future trends including the heavy use of IT, 24/7 production with three global shifts, lean permanent core staffing with significant out-sourcing, increased specialisation for small firms, increased computer literacy, need to demonstrate an understanding for human behaviour and lifestyles and the understanding of social roles and economic implications. These identified trends led CERF to a series of key questions on how these would bear implications for the built environments, including:

- Engineering emphasis: sustainability and the question of balance between economic, environmental and social imperatives, need for more global understanding and shift towards being a service industry, the age-old question of doing more with less and the possibility of reverting to the 'Master Builder' concept;
- Clients: the changing role of public agencies, large firms becoming major clients for small firms, growth of the non-governmental organisation (NGO) sector;
- Characteristics of projects: increase collaborative working, rise in prefabrication and mass production, growth of build-operate-transfer (BOT) sector, increase automation;
- Internet and software development: growth in the use of sensory devices, self-heal materials, use of technology reducing inspection and maintenance costs, electronic networking and data management;
- Workforce: increasing diversity amongst the workforce, emphasis on high-tech nature of the industry, move away from research-oriented to practice-oriented;
- Public relations for the professions: changing professional roles and liability, rising importance of risk management;
- Small firms: increased specialisation, delivery on request, more networking and more consolidation with larger firms;
- Miscellaneous: decentralisation of infrastructure, increased knowledge of advanced materials, understanding of the interaction of energy, information and infrastructure.

The American Society of Civil Engineers, USA (ASCE)

The ASCE was founded in 1852 to represent members of the civil engineering profession globally. To date, the ASCE has around 150 000 members worldwide and its vision is to position engineers as global leaders in a quest to build a better quality of life. A summit was organised by the Association over 2 days in June 2006 to produce *The vision for civil engineering in 2025.*[15] Around 100 participants comprising civil engineers, engineers from other disciplines, architects, educations and leaders from government, institutions and business participated in this visioning exercise. Of particular concern at this summit was the contribution made by civil engineers to the well being of society and the natural environment when designing and constructing physical infrastructure. The environmental agenda and social responsibility framed in terms of public health, safety and welfare certainly featured prominently as a backdrop to the discussions at the summit. A number of critical issues for the future were identified by summit participants, which form the basis for an action plan created by the Society in August 2009.[16] These issues included:

- The pressing need to embrace sustainability;
- The impacts of globalisation on engineering practice and the need to attract the best and brightest to the profession;
- Increasing demands for energy, drinking water, clean air, safe waste disposal and transportation;
- Greater need for collaborations forged across disciplinary and professional boundaries;
- Increasing need to engage with research and development, especially given technological developments in information technology, intelligent infrastructure, digital simulation and nanoscience;
- Better understanding of risk management especially in the age of uncertainty characterised by natural disasters, security threats and public finance; and
- Changing demands of clients and owners.

Chalmers University of Technology, Sweden

Chalmers University is based in Gothenburg, Sweden. In 2000, a team of academics from both Sweden and the UK produced a *Vision 2020*[17] document in response to a think tank that wanted to know more about how technologies over the next 20 years would impact on Sweden's construction industry. The vision was generated after an extensive literature review and consultation with the industry via workshops. Three key future trends were identified: globalisation, 24-hour operational capability through the 'virtual workplace', and pollution, global warming and environmental issues. These led to a number of recommended areas of emphasis, including:

- New and smart materials: growth in the development and use of materials that self-heal and are adaptable to the environment;
- Biomimetics: increased exploitation of technology that has the ability to mimic how nature deals with problems of adhesion, keeping warm (or cool), etc.;
- Nanotechnology: nanotechnology and the implications of material technology;
- Embedded systems: increased use of such systems to control, monitor and assist the operation of equipment/machinery/plant, use of such systems to enable/enhance communications;
- E-business: diversification and increased use of the web (and its different forms);
- Human capital: increased need for lifelong learning, recruitment and retention of younger workers, evolution of educational programme to allow people to retrain themselves and increased job mobility, and use of virtual reality to teach.

The Cooperative Research Centre, Australia (CRC)

The CRC for Construction Innovation was formed in 2001 as an Australian national research, development and implementation centre focused on the needs of the property, design, construction and facilities management sectors. Emulating global efforts into establishing future scenarios, *Construction 2020*[18] commenced in November 2003 to capture the Australian industrialists' perceptions of future scenarios, with a view to identify drivers and barriers that would respectively enable and inhibit the future, and to establish research gaps that needed to be plugged into. Through a series of exploratory workshops held in every capital city of Australia between November 2003 and February 2004, a questionnaire was produced to extract the current situation and the practitioners' views of the future. Validation workshops then proceeded to clarify and confirm the findings and to gain further support from the industry and agree on future action points. Eight headline visions resulted from the exercise, which are summarised as follows:

- Environmentally sustainable construction: climate change and depletion of natural resources, need to look for alternative energy sources, need to consider triple-bottom-line accounting (social, economic and environmental), education for environmental awareness;
- Meeting client needs: flexibility in design, improved client requirements capture, concept of more informed client;
- Improved business environment: increasing importance of alliances and collaboration (within supply chain and between industry and academia), use of ICTs to enable improved communication;
- Welfare and improvement of the labour force: growing computer literacy, increased dynamism in developing the workforce, improvements in health and safety through training;
- ICTs for construction: increased reliance on mobile technology, improved capability through training;
- Virtual prototyping for design, manufacture and operation: improvements in virtual reality, growth of 'try before you buy' concept;
- Off-site manufacture: increasing need to focus on off-site manufacturing, need to consider economic, social and environmental benefits;
- Improved process of manufacture of constructed products: supply chain integration and development of an industrial process protocol.

Synthesis of critical trends and implications from the foresight reports

From a political perspective, virtually all the 'futures' reports recognised the trend of globalisation and how this demands the need for greater collaboration. At the same time, it is interesting to note the expectation of increasing decentralisation of government and devolution of power to the regions and localities. Given the major role that the government plays in the industry, this would have implications in terms of how public spending policies are concocted. Certainly, up until the global financial crisis, relative freedom of capital movement has seen the rise of private equity finance, and the proliferation of schemes such as private finance initiatives (PFI), as the guardians of the public purse develop more innovative ways of funding infrastructure development.

Socially, globalisation sees the intensification of migration. The enlargement of the EU, for instance, implies greater (freer) movement of labour across member states, bringing with it the challenge from a human resource management perspective. Of course, the changing demographics also result in changes in consumer tastes, which necessitate consideration when producing design and construction solutions for the evolving client. In this respect, the reports have pointed to the trend towards greater individualism manifested in smaller sized households and the bourgeoning need for constructing more affordable single-person dwellings. In the age of consumerism, the role of clients and end-users will also gain increasing acknowledgement by construction professionals. From an economic perspective, business performance measurements and management will take into account more intangible forms to consider issues like customer satisfaction. In raising the professionalism of the industry, greater credence will be accorded to the role of the knowledge worker.

Technologically, ICTs were considered by many foresight studies to play a greater, more significant role in construction, which has traditionally been a slow adopter of innovation. Visualisation of the design and construction process is likely to proliferate. Enhancement of technologies is likely to be influenced by growth in the nanotechnology sector. There are also possibilities raised about the use of more automation and rapid prototyping technologies that will see less reliance on workers to physically construct buildings. However, counter-arguments have also been raised in some of the foresight reports that present a less sanguine view of such advancements, as calls were made for the need to maintain a more balanced perspective of the social implications of technological progress.

The climate change agenda was also forecasted to gain more prominence as debates about energy consumption and alternative sources strengthen. It is likely that greater restrictions will be placed on the extraction, production and consumption of building products, reflected in changes to building regulations. Solutions will also have to be found to address the impacts of climate change, e.g. increased flooding and disaster and crisis management. This will probably result in stricter planning rules regarding the use of greenfield sites. Furthermore, the role that professionals play in the future and the way the behaviour of built environment professionals is being regulated have been called into question.

Tables 1.3, 1.4 and 1.5[19] illustrate the emerging trends and implications for the built environment as recommended in the foresight reports reviewed in this section.

Critical appraisal of foresight studies

It is interesting to observe how the various reports reviewed in this section have been somewhat consistent in discussing the emerging themes that are critical for securing the future of the construction industry. In many respects, there is a great deal of convergence in the agendas raised in the foresight studies and the past reports originating from numerous post-war inquiries mentioned above. As the review highlighted, the reports dealt essentially with changes faced by society across political, economic, social, techno-logical, legislative and environment dimensions at a particular point in time. The reports have also mostly been compiled through consultation with major stakeholders of the industry, typically with representation from client groups, regulators, professionals and academics. However, as the nature of stakeholders changes, and as power relations shift from one stakeholder to another, there is undoubtedly divergence as to whose views get represented in the reports as the constituents alter over time. This does influence particular

Table 1.3 Summary of key trends extracted from the selected reports

Trend	nCRISP	CIRIA	CABE	CII	CERF	ASCE	Chalmers University	CRC
Ageing population	●●●	●						
Flexible working and living	●●		●●		●		●	●
Rise of the individual	●●	●●					●	●
Globalisation		●●●	●●●	●	●●●	●●●	●●●	●●●
Move to service industry	●●	●●●	●●●	●	●●●	●	●●●	●
Increased use of ICT		●		●	●			
Demand for lifelong learning							●	
Sensors and communication technology		●	●	●			●●●	
Automation						●●●	●●●●	
Nanotechology	●●●	●●					●●●●	●●
Climate change	●●●		●					
Alternative energy sources		●●			●			

ASCE, American Society of Civil Engineers; CABE, Commission for Architecture and the Built Environment; CERF, Civil Engineering Research Foundation; CII, Construction Industry Institute; CIRIA, Construction Industry Research and Information Association; CRC, Cooperative Research Centre; ICT, information and communications technology; nCRISP, Construction Research and Innovation Strategy Profile.

Table 1.4 Summary of the key implications extracted from the selected reports

Implications	nCRISP	CIRIA	CABE	CII	CERF	ASCE	Chalmers University	CRC
Smart materials and buildings	●●	●●		●●●	●●●●●	●●	●●	●●
Sustainability agenda	●●	●●	●●●●	●●	●●●●●	●●	●●	●●
Prefabrication and standardisation	●							
Mass customisation								
E-everything	●	●	●	●●	●●●●	●	●●	●
On-site automation							●	●
Customer focus		●	●	●	●	●●		●
Housebuilding/infrastructure		●●			●			
PFI/PPP		●	●●	●	●	●●		●
Self-build	●	●						
Refurbishment/renewal	●●	●	●					
Planning restrictions	●●	●		●●●●	●●	●●		
Global competition	●●	●●	●●	●●●	●	●●●		
Invest in People	●	●	●	●●●	●		●	●
Growing worth of professional judgement								

ASCE, American Society of Civil Engineers; CABE, Commission for Architecture and the Built Environment; CERF, Civil Engineering Research Foundation; CII, Construction Industry Institute; CIRIA, Construction Industry Research and Information Association; CRC, Cooperative Research Centre; nCRISP, Construction Research and Innovation Strategy Profile; PFI/PPP, Private Finance Initiative/Public–Private Partnership.

Table 1.5 Drivers for change in the future

Drivers	Critical issues	Main threats
Ageing population	• Workforce capacity issues • Changing consumer profile • The use of technology and automation as substitutes • The employment of immigrant labour • Public image of the industry	• Lack of preparedness • Sustainability of finance (e.g. pension funds)
Climate change	• Increased legislation – Cost, knowledge and training • Transformation of building regulations • Innovation in technologies (e.g. under-water construction)	• Cost and political will • Resistance to technological advancement

Individualism	• More single person accommodation
	  – Affordability
	  – Location
	  – Newbuild versus refurbishment
	• Increased traffic
	  – Impacts on the environment
	  – Transport infrastructure development
	• Increased leisure and support facilities
	• Importance of education
		• Cost
		• Politics/funding
		• Uncertainty over the balance of public/private provision
		• Planning issues
		• Difficulties to change human behaviour

Lifestyle expectations	• More and better products • More and better paid jobs • Better quality products • Balance of choices and opportunities	• Widening gap between rich and poor • Money, savings and lack of education
Political change	• Europeanism • UK in charge of own destiny through local/regional government	• Infighting/bureaucracy
Rise of China/India	• Opportunity to export our expertise • Material shortages • Capitalising on immigrant populations	• Lack of awareness and understanding of markets
Technology and communication	• Companies need to invest in research and development to stay ahead • Skills and training • Businesses require smart buildings • New entrants into construction market • Industry growth	• Lack of market stability for investment • Uncertainty of benefits from research and development • Short-term culture
Terrorist threat	• Surveillance • New markets	• Fear

emphases placed on each foresight report. That said, the constitution of the committees that produce these foresight reports is such that those involved are likely to represent an elite few in government, industry and academia. Therefore, recommendations tend to focus on high-level strategic policy matters with relatively less emphasis on how the implementation might be realised.

It is notable that the foresight reports tend to deal with the tensions arising from socio-technical change when discussing implications of future trends. On the one hand, the response of many foresight reports remains incredibly positive about the promises of greater efficiencies and reduction of physical labour made by technological progress. On the other hand, there is also recognition of the social implications of future scenarios, although the impacts on the livelihoods of those who work in the industry are often downplayed. Any acknowledgement of social impacts tends to be framed in relation to the market, usually in terms of changes in consumption patterns. Another tension that is inherent in many foresight reports is the issue of time-frame. The foresight reports reviewed in this chapter all discussed future trends and implications within a 15–20 year horizon. As we have argued above, because foresight reports are really about grasping with societal change in the world we live in today, futures thinking tends to be an extension of present-day thinking. This is certainly the point made by Harty and colleagues,[20] who argued that many foresight studies tend to focus on the matters of the day, as they questioned whether these reports really generate a strategic, long-term perspective of how the world we live in today can be revolutionised.

At best, foresight reports present commentaries of present-day challenges confronting society, with a view to offering alternative perspectives – framed as future scenarios – to tackle these challenges. The foresight reports tended to be compiled by committee, and, therefore, lacked the personalised view of how such 'futures thinking' can be enacted. Such institutionalised accounts of the future are often divorced from those working at the coalface of operational realities at the grassroots level.[21] Indeed, it is not clear how the recommendations of various foresight reports are implemented beyond the rhetorical level. It is our suspicion that foresight reports are just simply crystal-balls for future gazing; it is probably difficult, and indeed a futile exercise, to figure out what real action exactly derives from which report. If foresight studies were to realise their intentions of engendering change in industry and society, there is a pressing need to personalise 'futures thinking'.

In this book, we ask the fundamental questions of what exactly is 'futures thinking' in the construction, and more critically, how is 'futures thinking' shaped by the individuals involved in such an exercise? In answering these questions, it is the intention of this book to seek explanations for the process by which future scenarios for the construction industry are framed, and to provide greater clarity as to how much of the thinking about the future is derived from the occurrences of today and legacies of the past. We have also observed that many foresight reports tend to present an objectivised view of what the future might look like, and rarely explain how such thinking is influenced by those involved in its generation. Therefore, a chief objective of this book is to articulate how individuals who are likely to participate in such a 'futures thinking' exercise help formulate the outcomes through their personal value systems.

A note on the methodology

The idea for this book originated in early 2004 when we were interested to find out what trendsetters in the UK construction industry thought about the future of society and how this impacted upon the work of the industry. We drew particular methodological inspiration from *The change makers*[22] and *Business elites*,[23] in which Cooper and colleagues investigated the motivations that drive corporate leaders in the UK and how these drivers help influence the nature of British businesses and industry. So, by the end of 2004, we set out to formally invite a wide spectrum of high-profile participants from a range of organisations including designers, contractors, government officials and client representatives, by writing to them to explain our intentions. A total of 15 people (see interviewee biographies in the appendix; and Table 1.6) responded positively to this call, and their interviews form a large part of the material presented in this book.

From middle of 2005 to the end of 2006, we undertook the interviewing process. The pressures of finding a mutually convenient time to meet with our often very busy participants meant that there was inevitably a longer gestation period before data for this project could be amassed for analysis. Each interview was nevertheless done consistently. Three key questions form the basis for the interviews. In the first instance, we were interested in gaining a deeper understanding of the interviewee's personal and career journeys, including the personal and career issues that were critical influences that led them to where they are today. Second, we wanted to find out our interviewees' thoughts about the main challenges confronting the industry today, and how these agendas have altered throughout their lifetime. Third, the interviews were designed to capture their

Table 1.6 Brief description of the participants interviewed

Participant	Organisation at time of interview	Role
Alan Ritchie	Union of Construction, Allied Trades and Technicians (UCATT)	Trade unionist
Bob White	The MACE Group	Contractor
Chris Blythe	Chartered Institute of Building	Professional Institution
Chris Luebkeman	Arup	Engineer
George Ferguson	Royal Institute of British Architects	Architect
Guy Hazelhurst	ConstructionSkills	Government
Jon Rouse	The Housing Corporation	Government
Kenneth Yeang	Llewelyn Davies Yeang	Architect
Kevin McCloud	*Grand Designs*	Media
Nick Raynsford	MP for the Labour Government	Government
Sandi RhysJones	The Simon Group	Contractor
Sir Michael Latham	Construction Industry Training Board (CITB)	Government
Stef Stefanou	John Doyle Group	Contractor
Tom Bloxham	Urban Splash	Developer
Wayne Hemingway	Taylor Wimpey	Designer

Table 1.7 Interview protocol

Category	Key questions
Personal and career history	• Describe for us the career path you have taken. How did you come to work in the construction industry? • How has your personal life (from the earliest days you can remember) influenced the decisions you have taken in your career choices? • What were some of the critical events in your personal and career lifespan so far? Why were these critical? And how did these influence what you did? • Who are the biggest influences in your personal and career lifespan?
Present issues	• In your opinion, what are some of the present day issues that are confronting the construction industry? Why are these critical? • How have the emphases of these issues changed over your personal and career lifespan so far? • How have you dealt with some of these issues in the past, and even the present?
Future challenges	• What, in your view, are the big issues of tomorrow? • What, in your view, should we be starting to think about or do to address these future challenges?

It must be noted that although there is a clear structure presented here, the interviews were often carried out in an open conversation manner, and so the distinction between the three categories is more fluid than outlined.

perspectives of the big issues facing the industry in the future. A semi-structured protocol was used to help facilitate the interview process. This protocol is summarised in Table 1.7.

Each interview lasted up to 3 hours. Consent was given by the interviewees to record the proceedings and these were transcribed verbatim, yielding a total of around 150 000 words, for textual analysis. Additionally, cross-references were made to validate some of the assertions made during the interviews. This entailed inter alia checking through available records of company information and searching through news archives where appropriate. The text derived mainly from the interviews was analysed thoroughly and iteratively to identify emerging themes. Determining what these themes should be was an arduous task, in part due to the unwieldy nature of such qualitative data. Nonetheless, references were also made to the themes emanating from the foresight reports, as well as a more comprehensive review of the literature surrounding emerging issues. Of course, the conclusions presented in this book are solely our interpretations; therefore, we have attempted to re-present at great length as much direct quotation as possible, so that readers can make their own minds up on whether our conclusions hold up to scrutiny.[24]

At this point of explaining the methodology, it is critical that a main caveat is stated. For those readers used to reading standard reporting of research findings by Ph.D.-trained researchers, who diligently adopt a rational, scientific process, this will not be the case here. To all intents and purposes this book is not designed to report the results of a conventional research project. Instead, just as we were motivated to embark on this journey ourselves in 2004, we hope that this book will provide some fresh insights into the way influential people, who have elected to participate in this endeavour, have struggled through their personal and career lives to frame their thoughts about future challenges facing the construction industry. Although we do not lay claim to putting forward a representative view of the critical issues, we have sought to ensure that as broad a range of people as possible was consulted as part of the data collection process. We must, nonetheless, state a critical contribution of this book.[25] By tracing the intimate details of how our chosen participants have gone through their personal and career lives connected with the construction industry, we have therefore employed a rigorous and robust methodology that is rarely deployed in the field of construction management, i.e. life histories.[26] Of course, the validity and reliability of the findings remain a critical cause for our concern. More basic research might be required to verify the truths contained in our interviewees' perspectives. However, it is maintained that for the purpose of this book – that is to elicit the personal motivations and value systems of those who tend to be involved in shaping futures thinking in construction – the personal life stories of our participants should suffice. We hope you enjoy mulling over the stories of our participants and that this book will stimulate more scholarship into better comprehension of how we shape our understanding of the future challenges facing the construction industry.

The structure of the book

This book is split into three main parts.

Part 1: Tracing the past

In Chapters 1 and 2, we focus our attention on the past. In this chapter, we have reviewed how agendas of the past connect with various foresight studies written at the turn of the twenty-first century that help identify trends of the future that might impact on the construction industry. In Chapter 2, we will review the theoretical concept of leadership in construction alongside an analysis of the stories of leadership development by the influential figures interviewed for this project. The findings reveal a rich picture of where the leading figures have come from in their personal and career lives thus far, how they have developed in thinking and practice, and how these contribute to their thoughts about ongoing and future developments in the construction industry. The conventional idea that leadership is a tangible thing that can be easily defined and measured through a set of attributes is being challenged, and an argument is put forward to view leadership as more of an emergent process.

Part 2: Eliciting the future

In Chapters 3 and 4, we focus on the analysis of our interviewees' thoughts about future challenges that the industry is attempting to grasp in the present day. Chapter 3 will focus on the issue of sustainable development. Sustainability emerged as one of the most important agendas for the leading figures interviewed. The analysis of their thoughts reveals an acceptance of the need to trade-off between economic, social and environmental concerns, and that skills development is needed to ensure the pursuit of a sustainable future. An attempt will also be made to compare the perspectives of our interviewees with a review of literature based on various theoretical perspectives of sustainable development of how the construction industry can contribute to sustainable development. Whereas Chapter 3 stresses the importance of human agency and relations in meeting the sustainable development agenda, Chapter 4 looks at the concept of governance with a view to appreciate how an institutionally coordinated response to sustainable development can be achieved. Positive action requires strong governance. Chapter 4 traces our interviewees' thoughts on a wide range of key influences, including globalisation, the intensification of public–private collaborations, the changing organisational forms of construction firms, the role of education and research and the future meaning of professionalism in the construction industry. These thoughts correspond closely with the general literature on governance, which traces the paradigmatic shifts that evolve in terms of overlapping political, corporate and community levels of governance.

Part 3: Towards an afterthought

In the final chapter, we conclude with the key lessons learnt by pulling together the various strands discussed in the preceding chapters. This chapter will focus on three fundamental messages:

- First, we established that 'futures thinking' is really about emergent thinking, by which policy-makers make sense of the complexities of socio-technical change;
- Second, societal change disrupts the established order of doing things and constantly dismantles boundaries of the past. Such disruptions contribute to the impetus behind efforts to engender 'futures thinking';
- Third, although the removal of boundaries brings about the promise of exciting, new ways of doing things, it also creates a number of tensions and paradoxes that practitioners have to contend with. Articulating 'futures thinking' is a way in which practitioners make sense of these tensions and paradoxes. The chapter calls for more scholarship to understand how these tensions and paradoxes are resolved by practitioners.

On the whole, this book is about understanding how societal trends of the future are being shaped by those who engage in 'futures thinking'. Through scrutiny of the value systems of influential figures in the UK construction industry, it is the intention of this book to articulate how these contribute to more effective ways of framing an agenda for achieving a more sustainable future. It is hoped that you will enjoy reading this, and, like us, benefit from understanding how 'futures thinking' is derived.

Chapter 2

Influential people in the UK construction industry: what makes leadership in construction?

'A life spent making mistakes is not only more honorable but more useful than a life spent in doing nothing.'

George Bernard Shaw, 1856–1950

Chapter summary

In the previous chapter, it was argued that research and policy efforts to report on future scenarios planning are often divorced from a deep exposition of the value systems of those who frame such future agendas. To gain a better understanding of the future development of the construction industry, it is vital for us to gain insights about the people who 'lead' the industry. This chapter is, therefore, dedicated to offering an explanation of how our interviewees, chosen as leading figures representing the UK construction industry, have arrived at the position of influence in the present day. To do so, the chapter comprises a salient review of mainstream leadership theories, as well as studies undertaken specifically to analyse leadership in construction. Tales of how our interviewees developed – both personally and professionally – are then re-presented to make sense of the complexities and dynamics of the workings of leadership in the context of construction.

A review of the leadership literature reveals that the understanding of construction leadership is somewhat primitive, failing to consider the relatively mature developments of mainstream leadership theories. Furthermore, mainstream scholars raise the need to examine the context of leadership as part of broader sociological, historical and political developments, rather than simply reinforcing the ad nauseum emphasis on managerial functionalism. Our analysis of the interviews illustrates how the critical leadership antecedents of people, places and events help shape the thinking of our leaders. Furthermore, we discuss how a number of our interviewees tend to take on the role of starters rather than finishers. Such desire to move on from one idea to the next probably accounts for the fact that these leaders tend to be connected to a wide range of people as they embark on a lifelong journey of learning. However, there is nothing prescriptive here about the nature of leadership, as the label itself is rather less meaningful than what our

Constructing Futures: Industry Leaders and Futures Thinking in Construction   Paul Chan and Rachel Cooper
© 2011 Paul Chan and Rachel Cooper

interviewees, in their positions of influence, actually do. In this chapter, we discuss how many of our leading figures have the tenacity to want to make a difference for the betterment of the industry; that they are often willing to seize every opportunity to make an impact and influence an outcome. We will also note how our interviewees are purveyors of the establishment on the one hand, yet, on the other, they are also keen advocates of change within the institutions in which they serve.

The key issues discussed in this chapter are as follows:

- People, places and events matter in shaping of the thoughts and practices of leaders in the construction industry. It is the range of experiences – both past and present – that continually define how leaders respond to the challenges of the day;
- Leadership cannot be defined as a static concept. Rather, leadership is an emergent process that emphasises the need for adaptation and learning to confront the ever-changing environment;
- The role of learning is critical in the enactment of leadership. Leaders are often avid learners of new knowledge, willing to step outside their zones of comfort and never afraid to learn from mistakes made in the past;
- Leaders often start from an esteemed position of wanting to make a difference in society. To be an effective leader, one really has to want to make an impact;
- There is much to be gained to emphasise the social dynamics of leadership development. Understanding life histories of influential people can be a useful start to help articulate personal agendas of those who are likely to shape the future of society.

Introduction

'It would be pleasing to think that the future was a blank screen on which we could design our future. The reality, as Ernest Hemingway once said, is that the seeds of our life are there from the beginning – if we bother to look'.[1] This book is about understanding how future scenarios can be shaped in the context of the construction industry, drawing from the perspectives of a number of influential figures in the UK industry. In this chapter, an understanding of what makes these people influential is being examined. We pose the question as to what makes a leader in the construction industry; and, to seek answers to this, we cast an eye on the past lives of the 15 influential people involved in the study, so as to appreciate their value systems and how these might help influence their vision for the future.

The mainstream field of leadership is well researched; most leadership theories and models that are still being discussed today have evolved from the scholarly work of the 1950s and 1960s[2] into identifying characteristics that enable individuals to become leaders (often perceived as people with authority) or to display leadership techniques.[3] Over the last decade, there has been a resurgence of interest in the study of leadership, in part due to wider recognition of the dynamic, changing business environment and the need for organisations to constantly adapt and innovate to survive. Popular writers[4] have talked of the need for new types of leaders to champion the competitiveness agenda based on innovation.

Despite the wealth of knowledge built around the concept of leadership, there are scholars who contend that the understanding of leadership has not been fully developed.[5] As this chapter unfolds, it is suggested that the application of leadership studies in the realm of construction management remains somewhat primitive. The contribution of this chapter, therefore, is to broaden the application of leadership theory in construction. The chapter is organised as follows: a review of the salient points of leadership literature is first presented, which reveals a need for a deeper examination of the values and belief systems that shape the development of leaders, before outlining what, in our view, makes a leader in construction based on the interviews undertaken.

Leadership theories and application in construction

'The practice and study of leadership has been, is, and will be a continuing fascination for leaders and academics.'[6] However, leadership is not necessarily found only in the elite few of the upper echelons in business organisations.[7] In a systematic review published by the Advanced Institute of Management Research (AIM) in the UK, Munshi and colleagues maintained that 'leaders are important at all levels in the organisation';[8] they considered leaders to perform two key roles: that of motivating others into uncharted terrains,[9] and to design organisational systems that enable employees to be innovative. In this respect, the concept of leadership is different from the study of management.

Several writers have considered this distinction. Kotter,[10] for instance, suggested that management is more about coping with the ordinary run of the mill in organisations, whereas leadership is about coping with change. In a similar vein, Grint contrasted: 'the division between Management and Leadership, rooted in the distinction between known and unknown, belies the complexity of the relationship between problem and response. Oftentimes the simple experience of *déjà vu* does not lend itself to the application of a tried and tested process because it is really "*déjà vu* all over again"'.[11] Fairholm put it simply, 'if you can count it, you can control it, you can program it, and therefore, you can manage it. If you cannot count it, you have to do leadership.'[12]

Nonetheless, Fairholm suggested that 'some still may not see a distinction'.[13] Indeed, critics have argued that the blurring of boundaries between leadership and management is due to the dominance of functionalism in management research. For Berry and colleagues, throughout 'the early development of leadership theories [. . .] the ontology was realist and the epistemology was functionalist [. . .] these also fit with the managerial functionalism of Henri Fayol'.[14] They added, 'The criterion for effectiveness is still the functional effectiveness of the leader's behaviours [. . .] yet Fayolian functionalism and structural functionalism have been in critical retreat for decades'.[15] At this point, it is, therefore, pertinent to trace the development of contemporary leadership theories and to examine their adoption into construction management research.

Development of leadership theories

Five clusters of leadership theories have been suggested in the literature:[16] traits and styles; contingency; transformational/transactional; distributed; and structuralist leadership theories. Early leadership theorists have been concerned with discovering the traits and styles that leaders possess,[17] believing that these characteristics are what differentiates leaders from followers.

However, sceptics have indicated that these characteristics do little to predict whether people who possess such traits will necessarily become leaders.[18] Indeed, as Fairholm stated, 'studying individual leaders may not get you to a general understanding of leadership [...] leadership is something larger than the leader – that leadership encompasses all there is that defines who a leader may be.'[19] Subsequently, a strand of leadership theories emerged that considered the context in which leadership is practised. Proponents of the contingency approach[20] subscribe to the view that any leadership response is dependent on the particular situation that warrants that response.

Contemporarily, the contingency approach has also received some criticism. Grint, for example, argued that 'the difficulty of separating the situation from the leaders [is] because the former is often the consequence of the latter',[21] rather than the other way round. Grint (2005) used the analogy of the Trojan Horse to illustrate this by claiming that 'the appearance of the wooden horse outside the walls of Troy did not require the Trojans to bring the horse inside the wall, they chose to do it', and added, 'This reassertion of the role of choice in the hands of leaders does not imply [...] they are determined in the actions by the situations they find themselves in.'[22]

Culture and power relations often run alongside the study of leadership. Burns[23] examined the leader–follower relationship, and differentiated between transactional and transformational leadership. The former relates to the rewards–punishment ('carrot-and-stick') framework that influences the behaviour of followers; and the latter points towards the emphasis on the effective articulation and communication of vision through such attributes as charisma.[24] However, others[25] have observed that analysing the effectiveness of transformational leadership is virtually impossible due to the subjectivities involved in personal styles.

Questioning the conventional leader–follower dichotomy, distributed leadership theorists[26] have suggested that new forms of work organisation have resulted in greater interdependence and coordination, which in turn give rise to the need for more distributed leadership practice. Using examples from the education sector, Spillane[27] objected to the often taken-for-granted view that those who work at the lower levels of the organisational hierarchy are necessarily subordinated to the leaders of their organisation; the logic that those who lead in organisations can expect those who work at the grassroots to follow, willingly or otherwise, remains questionable. Furthermore, given the trend towards more self-management and flatter business organisations, distributed leadership theorists insist that there needs to be consideration of leadership at all levels of the organisation.[28]

Organisational systems feature prominently in structuralist leadership theories. Popular writers like Peter Senge[29] talk of the need for leaders to engage in systems thinking when designing organisational learning environments to deal with the ever-changing business climate. Huff and Möslein[30] suggest that a crucial role of leaders is to design organisational structures that facilitate effective distribution of resources. Leaders fulfil the role of 'architects in an administrative sense'.[31] However, structuralist leadership theorists have begun to consider the dynamic interactions of the people who are being subjected to the organisational systems designed by 'leaders', echoing the long-standing sociological debate on duality of structure and human agency. For example, Collinson[32] draws on the work of scholars like Giddens[33] to consider interdependent relationships between leaders and followers, contending that 'followers' practices are frequently proactive, knowledge-able and oppositional [...] and that leaders themselves may engage in workplace dissent',[34] reinforcing shifting power relations that align with the distributed leadership perspective.

Application of leadership theories in construction management research

The adoption of leadership theories in construction has been somewhat primitive when compared with the relatively mature development of mainstream leadership literature summarised in the preceding subsection. Many leadership studies in the field of con-struction management research have been concerned with merely examining leadership effectiveness in relation to organisational performance. For example, Odusami and colleagues,[35] through analysing data collected from 60 questionnaire surveys from project leaders in various professions, investigated the relationship between project leadership and construction project performance. The attributes of project leadership found in their survey instrument originated from the four leadership styles (shareholder, autocrat, consensus and consultative) developed by Slevin and Pinto.[36]

In understanding the role of leadership in promoting construction innovation, Nam and Tatum[37] interviewed more than 90 construction professionals involved in 10 innovative projects that took place in the USA in the late 1980s, and concluded that effective leadership implied the need for leaders who are technically competent entre-preneurs who can drive forward innovation. In a similar vein, McCabe and colleagues[38] examined the nature of leadership in the management of quality. These studies tend to support the traits and styles strand of leadership theories, and augment the emphasis on managerial functionalism observed by Berry and Cartwright.[39]

Of course, culture plays an important role in the study of construction leadership. For example, Low[40] contrasted between Eastern and Western philosophies to discuss the relevance of the teachings of Chinese philosophy, Lao Tzu, in construction project leadership. Fellows and colleagues[41] also investigated leadership styles and power relations in quantity surveying in Hong Kong. Other behavioural research into construction leadership includes Dainty, Bryman and Price,[42] who discussed the essence of leadership in empowerment within the UK construction sector.

There is no doubt that these studies provide illuminating insights into leadership in construction. However, the emphasis remains narrowly focused on the performance agenda and does not consider progress made in the trajectory of mainstream leadership theories outlined above. There is indeed a need to move forward with examination of leadership in construction and this is the intention of the work reported in this chapter. The next section will highlight some of the emerging issues to take this forward.

Looking forward but learning from the past

The observation provided by Berry and Cartwright that research on leadership has hitherto concentrated narrowly on managerial functionalism and the effectiveness agenda is indeed true for studies undertaken in the area of leadership in construction. In fact, Fairholm argues more forcefully, 'Researchers have denigrated the idea of leadership [. . .] because they misunderstand the evolving nature of authority derived from changing social structures, and because they have missed opportunities to tie in research procedures and focuses from intellectual interests such as psychology, sociology, history and political science, not just scientific management, Weberian bureaucracy, and the like'.[43]

Berry and Cartwright pose a further question, 'How then do persons become leaders?' and add, 'From early constructions that leaders were born and schooled in a given social class via constructions of entrepreneurs as leaders, the literature has been opaque upon the actual process of leader formation'.[44] Indeed, Cooper and colleagues[45] suggested that before one looks into developing authentic leaders, one needs to learn from the past of the individual. As Samuel Taylor Coleridge beautifully penned, 'If men could learn from history, what lessons it might teach us! But passion and party blind our eyes, and the light which experience gives is a lantern on the stern, which shines only on the waves behind us!'. This is what this chapter seeks to do by examining the past lives of leaders in UK construction to better understand underlying values and belief systems that influence their current practice and thoughts on the future.

So what makes a 'leader' in construction?

This section presents the findings from the analysis of the past lives of our leaders. The analysis reveals the critical leadership antecedents of people, places and event that shape the thinking of our influential participants. In essence, the findings point to the age-old balance between nature and nurture, i.e. leaders emerge as a result of both the innate qualities they possess and the environment to which they have been exposed. These key points will now be elaborated below.

Critical antecedent: people

It is commonly accepted that the construction industry is extremely paternalistic and many writers have alluded to the power of family relations in encouraging new entrants into the industry.[46] So, it is unsurprising to find this same phenomenon manifested in our interviews. Many of our participants talk about how they were influenced by their family and social circles to get involved with the construction industry. For example, Chris Luebkeman spoke fondly of his grandfather, 'My grandfather was an inventor and he studied what we could call civil engineering in the broadest sense as well as geology. He probably had a big influence on my life'. He went on to explain his grandfather's influence, 'When I was a young lad, I worked as an apprentice carpenter on weekends and in the summertime for about five or six years. Afterwards, I went to university to study and got a double major in Geology and Civil Engineering at Vanderbilt University in Nashville Tennessee'.

Charles Handy reflected on his life and commented that the fact that his many family relations were teachers should have provided him an insight into his teaching career, suggesting that 'our past is inevitably part of our present and also of our future'.[47] It is perhaps not coincidental that Chris Luebkeman followed in the footsteps of his grand-father. George Ferguson also acknowledged the role his father played in encouraging him to become an architect, as he reflected, 'Although my father wasn't telling me to be an architect, I think he always showed a great interest in buildings and he could draw very well. He was an excellent craftsman too, especially joinery and cabinet making'. Alan Ritchie was also initiated into the trades when his father signed him up for an indentured apprenticeship.

The support of family members is indeed instrumental for encouraging new entrants into the construction industry, and this is particularly the case for minorities wanting to work in the industry. For Sandi Rhys Jones, she explained, 'I love the Built Environment, I am fascinated by buildings and big structures. I remember as a small child sitting and just watching cranes. I helped my father in practical jobs around the house. My father was an electrical engineer and was a very practical man and he taught me as a child how to saw wood, how to hammer nails. He taught me to paint and decorate. I enjoyed it. It was something tangible and that continues to be so. I am the painter and decorator in our family, I can French polish and upholster and I like general building work including bricklaying. I would like to come back in another life as a bridge builder'.

Family members can also shape the way we behave in society. For Stef Stefanou, he attributed his business acumen to his father, who had a lasting impact on how he manages the affairs of the construction business he chairs today, 'I can always remember my father. To him, the principle that the customer is always right applied. And the lesson I got out of that was that the client is the most important part of the company, because no matter how many good people you have, without a client, you will have no business to start with. The second thing is that really you have to look after your staff, be they the cleaners or be they the directors. Again, you can have the best clients on Earth, but if you haven't got good people, you're finished'. As we will see in the later chapters, Stef – and indeed all our interviewees – feels passionately about the significant role that human relations play in the effective delivery of work done in the construction industry.

Access to senior people

Notwithstanding the importance of people in shaping our interviewees, and the part that family members play in promoting the industry to the interviewees during their formative years, it is also apparent that the key to achieving success for many of our leaders was the access to senior people during the early stages of their career. This was the case for Chris Blythe, as he remembered one of his earliest work experiences during his placement in industry: 'I went back to college and my third placement was with Esso Petroleum in Birmingham. They had just opened a new terminal and I got involved in some interesting projects there. I worked on a project with a girl from Aston University in the regional head office for a while for the authorised distributor network in Worcestershire and we were asked to come up with some recommendations as to how that would work. We were only young undergraduates and we were given this exercise to do. We did the presentation to senior management and we implemented its proposal and looking back at it now, I'd be scared to death to do a project like that, but at the time, we did it and it was wonderful'.

This not only developed his confidence in dealing with senior people, but also remains his pride and joy today, 'If I'm ever up in the West Midlands, and I see a particular authorised distributors, I think, "Yes, we were behind that from the work I did back in 1975"'. Nonetheless, the ability to shine in front of senior management at that stage in Chris's life has certainly opened many subsequent doors for his future career.

For Bob White, his moment of epiphany was the opportunity to work on a high-profile project with senior people in Bovis. He explained, 'My mentor at Bovis then was Ian McPherson. He was then number 2 in a division and we got on very well. And then, when Broadgate came along, I decided I'd had enough of working in an engineering type role, so I said to Ian McPherson when Broadgate was on the horizon, "Look, if we win this job, I want to be a part of this thing". And he said, "Yeah, happily". And I mean, Broadgate was the first major development in terms of construction management and Bovis, as a result were well ahead of the pack in terms of applying that process. As a result of that, both Ian and I decided to set up Mace at the end of 1990'.

The importance of networking

The preparedness to network was evident in the way some of our leaders gained access to senior people in the industry. For example, Kevin McCloud, known for his television series *Grand Designs*, remembers his opportune moment in getting into television. 'It was because I had written a book about lighting and somebody said, "Oh, we need somebody to talk about lighting" around 1994. This was when, actually as a discipline, there were only two books on the subject and that was it, one by me and one by a girl I know very well, Sally Story. She was a much respected writer in her time. But anyway, they paid me 100 quid or something and that was it and then they asked me to do it again for another show and again and they kept saying, "Oh, Kevin, you're good. You're good on television. You've got to talk about this. You've got to talk about that, even though you don't know anything about it". And I would oblige'.

In a sense, this mirrors the way Sandi Rhys Jones got to lead the influential Equal Opportunities Working Group in 1996, which promoted inter alia one of her greatest missions in the industry – women and equality in construction. 'Sir Michael makes the point that there were very few women and this is madness. . .I managed to get an invitation to go to the launch of the Latham Report and to ask a question. They had the great and the good lined up and I said how pleased I was to see the emphasis on the realisation that there were very few women in the industry and that this should change. "What practically was going to be done about that and was there going to be greater represen-tation of women in the process?" I asked. At that point, there was a lot of shuffling and Sir Michael says, "Sandi Rhys Jones asked a question as to whether we were going to do anything". It was subsequently agreed that there would be a Working Group on Equal Opportunities and I was proposed as a member by the Association of Specialist Sub-Contractors. Six months after it had started, I became Chair of the Working Group – 16 men and women - and we focused on gender'. Therefore, the old adage of being at the right place and at the right time, and so it seemed saying the right things, mattered in securing a role for our interviewees to exercise their influence on the industry in their subsequent career lives.

The sum of the parts

In spite of the stories recounted by our interviewees, and specifically naming senior people who have had an influence on their lives, none of our interviewees considered that a single individual could claim credit for shaping what they do in their professional practice. For our leading figures, it was usually a wide range of people that had crossed their paths at various stages of their career, who collectively help mould their character and thoughts about their work. So, Alan Ritchie talked about how his late predecessor in the trade union was passionate about championing the employee voice. Bob White attributed to his early work experience with Henry Suede and Sid Bell from Nottinghamshire County Council shortly after graduation, when he was involved in an innovative project at that time called 'Research into Site Management (Project RSM)'. He explained, 'this was a remarkable learning experience, at a fairly young age, as an architect to actually design a building and then have to go on site and instruct the group of guys how to build your building that you designed'. Conversely, Chris Luebkeman recalls how his architect friends at university transformed his views about design beyond engineering drawings.

So, our influential people were themselves influenced by a variety of people whom they interacted with throughout their careers, and it is difficult to pinpoint precisely what causes them to think the way they do. Indeed, Nick Raynsford summed up the emergent, evolutionary process in which a range of people helped contribute to their thoughts over time, as he acknowledged 'A huge range of people, and there is no individual that I would say I'd learnt everything from or who'd inspired me most. I mean it's a very wide group of people, including great engineers, great architects, people who've headed construction, major construction contractors, people who have been inspired clients. There is a wide range it would be invidious to name individuals'. However, it is not everybody that can have that effect. Nick explained that it is 'the quality of the people that I have met' that mattered, and 'their commitment and dedication and ambition to work more effectively together rather than reverting to the old adversarial culture that sadly damaged the industry so much in the past' that have influenced the way he thinks about how the construction industry can develop in the future.

Critical antecedent: place

Geography also plays a crucial part in our leaders' thinking. This is natural given the physical and transient nature of construction work, where the essence of place plays a more prominent role when compared to other industries. So, Jon Rouse remembers his childhood spent in deprived parts of Bradford and Barnsley in West Yorkshire and his reception when his family moved to a relatively 'leafy suburb' part of Northamptonshire. 'When I was twelve or eleven, my parents mainly for the sake of mine, and my brother and sister's education, decided to move away from Barnsley and we moved to leafy, suburban Northamptonshire. The contrast was very stark, because when I was in Barnsley, although my dad was a social worker, you know, he wasn't earning a huge amount of money, and we were actually one of the wealthiest families. But when we moved to Northamptonshire, we nearly became the poorest family'. Jon explained that following his early exposure to income disparities, and having experienced the North–South divide, he became determined to work in the broad areas of housing and social regeneration to help combat poverty issues.

Tom Bloxham was inspired by the industrial landscape of Northern cities like Manchester, and again, the contrast between the dilapidation of buildings in Manchester in the 1980s and the relative vibrancy of places such as Camden Market in London motivated him to pursue a lifelong expedition to renew and refresh many of the derelict industrial buildings across the country. Tom stressed, 'Coming from Manchester and seeing an amazing stock of old buildings that seem to have been dumped because they were empty and, on every street there are boarded-up empty spaces. I had seen a potential for them. You'd imagine a load of really interesting designs, in terms of shops, maybe hairdressers, occasional restaurants, but nothing in terms of housing or offices. At the same time, I was aware of a load of people who were young entrepreneurs, who started selling clothes in Camden Market. You suddenly think there is an opportunity here'. It is again no accident that Tom still maintains an entrepreneurial drive when seeking out opportunities for property redevelopment in his pursuit to encourage mixed uses in buildings.

George Ferguson also explained how the city of Bristol instilled in him a passion for heritage preservation when he studied at University there, 'I found Bristol so fascinating. The university and the architectural school were bang in the middle of the city. I always thought that that was a very important part of my education. Being in a city and interfering myself with that city and taking an interest in it, I can't overestimate the contribution that being a citizen of Bristol made at that time'. As we shall see later on, living in Bristol as a student imbued a sense of political activism in George that would see his longstanding interest in the English heritage grow even more fervently. For Alan Ritchie, the sense of place did have a political dimension. Remembering his experience of delivering a speech to 100 000 UCATT (Union of Construction, Allied Trades and Technicians) members in Hyde Park, Alan Ritchie noted, 'I keep on saying to myself, if you want to be Prime Minister of this country, you couldn't do it in Paisley. So, you know, if you want to influence the politics of this country, if you want to be there at the top table of the General Council of the TUC (Trade Union Congress) you've got to go to London'.

The medieval term 'journeyman' was used to describe an apprentice who moved around with a master craftsman. In a sense, this is still very relevant to the world of construction. Indeed, our leaders have benefited from the experiences gained through travelling around. As with the critical antecedent 'people', our leaders certainly did not ascribe their development to a single place. As Chris Blythe suggested, 'I'm originally from Yorkshire. I was born in Bradford. I grew up on the East Coast. My father was in the Air Force, so we travelled around a lot. It's strange because I was away to boarding school and because my father travelled around a lot. So, I would say that all these influences probably came more by living in different schools and different homes and visiting different places when I was younger'.

Critical antecedent: events

Critical events during the personal and career lives of our leaders also contribute to the way they think about the world around them. So, Stef Stefanou, a Greek by upbringing who was raised in Egypt, considered the Suez Canal crisis in the late 1950s to be a major turning point in his life: 'You know the revolution in Egypt? Gamal Abdel Nasser in Egypt? When I left Egypt, the media kept saying that he was supposed to be a dictator who was suppressing all foreigners in Egypt at that time. But when I was in Egypt, I never saw people killing anybody. That was a major event for me. I used to spend hours and hours reading in Battersea Library what the English papers were writing about Egypt. I thought, "They must be writing from another country!" because I didn't experience it when I was living there. So, I started realising that there is a game here. I was an innocent kid that thought life goes on normally'. This event certainly helped make Stef Stefanou sceptical on any official reporting of current affairs, even to this day.

George Ferguson described one of his life-changing episodes while studying in Bristol, 'In my second year, I bought a house, which might sound extravagant. But I paid the deposit with the money I got from selling a patent on a game and the rest of it was bought with my student grant which was very small because the house cost £800 and the mortgage was £5 a week and I used to let part of it. Now, that may seem like an irrelevance. But, I wasn't just buying a house. It was threatened with demolition. So I was buying a campaign. I decided when I bought the house that what I was doing was to campaign for the future of this area'.

Chris Blythe's passion for developing people emanated ironically from his experience in making people redundant throughout his career, as he traced his first experience as an accountancy placement student, 'I had my first taste of insolvency practice. And that was a thoroughly unpleasant experience, you know. They were single-minded in what they had to do. For the staff that the firm took over, they were very tough and dismissive and I think, unnecessarily cruel'. He went on to add, 'A couple of years after that, funnily enough I went to work for a firm of accountants. And very early on, I was sent to do some work on a tannery down in South Wales. The aim was that the tannery was going to shut and it was going to go bust. You know, we thought it was going to go bankrupt. They had got all the redundancy letters because we were making about half of the staff redundant that day. But before we issued them, I got in touch with the local job centre and arranged for an official from the job centre to come to the place in the afternoon and as we were making people redundant, they would go to this job centre official and she then made an appointment for them to go into the job centre to make sure that they had all the forms to start a claim for benefit'. This demonstrates how early events had a lasting impression on Chris Blythe, and how he had learnt from first episodes of being involved in redundancies. This probably explains his later involvement in people development initiatives including Investors in People (IiP) and his current role as CEO of the Chartered Institute of Building (CIOB).

For Sandi Rhys Jones, it was a similar tale, as she described her experience as an executive officer at the British Post Office, 'Shall I tell you how I was known? C5/1. And you signed in to the minute and you signed out to the minute. I was defined at my desk as C5/1 and I can remember that now. It taught me was that you should never depersonalise people in the workplace and as I grew older, I realised quite how dreadful it was. By all means mark the desk with a number, but not the person. I learnt about managing people'. It is notable that Sandi Rhys Jones went on to dedicate most of her career to making everybody count in the industry, especially women and ethnic minorities.

For many of our leaders, however, critical events that shaped their professional thinking were often linked to a personal sense of achievement. Bob White, who co-founded MACE

Limited intimated, 'At Broadgate, I got my first exposure to construction management. And it was such a breath of fresh air to us. We all felt liberated because of the process. It all enabled us to perhaps utilise skills we thought we had much more in previous systems. It also allowed us to engage ourselves with the other partners in the supply chain process in ways we never did before. And we were re-engineered by it'.

Jon Rouse and Nick Raynsford both talked about the 'big break' in their working lives. For Jon Rouse, 'I had my big break. Basically, if it hadn't been for my big break, I don't think I'll be here today. I got a phone call from Richard Rogers and he basically said, "I've got a proposition for you. I've read your work. I've seen some of the things you've done and would you be willing to come and be Secretary of my Urban Taskforce?" and I spoke to Price WaterhouseCoopers and they released me on secondment for until after Christmas to produce *Towards an urban renaissance* which I ghosted. I wrote it with the oversight of the Urban Taskforce and I'm very pleased with that. Now I'm very proud of it. I don't think it's perfect by any means, but I think it was the basis of the shift in philosophy in the Labour Government, in terms of formulating the basis of their urban policy'. It certainly set Jon in good stead for heading up the Commission for Architecture and the Built Environment (CABE), then the Housing Corporation and now Croydon Borough Council.

For Nick Raynsford, 'The most important campaign that I was involved in during the early days was for the passage of the 1977 Homeless Persons Act, which was very much promoted by non-governmental organisations. Government was initially not certain it wanted to or not. Local government was generally hostile. So, that was quite a challenge to get the law changed. And we succeeded by means of a private members bill, which was given government backing, but it was nevertheless a private bill, so it wasn't a piece of government legislation. And that required a huge amount of lobbying from the organisations that I was working with to support the legislation'. It is no wonder that Nick is a keen advocate for seeking cross-partisan partnerships in politics and has often worked on the basis of forging collaborations, as we shall see in the later chapters. Nick was also instrumental in the decision-making process at the governmental level for the Channel Tunnel project in the UK, an event that first made him aware of the importance of the public image of construction.

To be or not to be? The age-old question of nature and nurture

The preceding section provided a rich picture of how people, places and events can critically shape the development of our leaders. This is based on our post hoc analysis of the career trajectories of our leaders. However, it does little to 'answer' the age-old question of nature and nurture, i.e. are leaders born or made? The remaining part of this section outlines our analysis, which suggests that it is probably the latter.

I started and so i'll finish?

It is interesting to observe that some of our interviewees display a preference for starting up projects, rather than completing them. In an industry typified by transience and the need to move on from project to project, the generation of ideas seems to appeal to our leaders more than the implementation. As Chris Luebkeman acknowledged, 'I've always worked, I've always worked more upfront. And I've known about how things are done, but I'm more of a starter than a completer. So, I enjoy that creative starting and trying to interpret and putting the pieces of the puzzle out on the table'. Our chosen leaders have all been starters in one form or another. So, George Ferguson started several campaigns in his life to preserve the landscape of English towns and cities; Bob White was so fascinated with construction management that he co-founded MACE Limited; Sandi Rhys Jones was instrumental in starting the women in construction agenda in the 1990s; Kevin McCloud, through his television programme *Grand Designs*, started a deeper public appreciation of the built environment; Wayne Hemingway and his wife, Geraldine Hemingway started inter alia fashion label *Red or Dead*, and so forth.

Jon Rouse also remarked, 'I've done two start-ups in my career. I've done the Energy Saving Trust and CABE and they are in many ways, the most exciting job you can do. To start something up from scratch, to shape it yourself, to build something new, I mean, it is obviously the most exciting thing you can do in your life. I'm a better starter than a finisher'. That said, Jon also suggested that 'the reason why I took on the Housing Corporation job is because I wanted to test myself in a normal turnaround situation where we had a city that's 50 years old with a lot of embedded cultural norms and attitudes, and one of the things I wanted to test was actually that proposition that I was good at taking an idea and running with it. You know, could I also have the discipline to actually work in a context where there was actually already a strongly embedded set of norms, culture, values and actually work with the grain, but at the same time revitalise?'. So, one could interpret Jon's desire to see something through as a way to seek a brand new challenge. It might be the case that the leaders we have chosen to interview are more inclined to be starters, rather than finishers. Yet, this extract from Jon suggests that they are also willing to go with a change of circumstances.

Kenneth Yeang certainly thinks that he oscillates between different roles, as he talked about how he likes both the generation and testing of ideas, but also seeing through implementation. He commented, 'You see, you have to develop the theory, you need to come up with the technical solutions and ideas and you have to test them out, and then you have to design the buildings. At the same time, you have to survive as an architect. So I sort of oscillate, if you like, between developing new ideas, trying to find out how they can work technically'. So, in an industry that is as diverse as construction, where opportunities abound in both ideas generation and implementation, the challenge of seeing through ideas is certainly tempting for our interviewees. For instance, Chris Luebkeman recalls the time when he was invited as an external assessor of a degree programme in Hong Kong and was latterly invited to implement some of his recommendations, 'We made recommendations and then about a month later the question came back and I said, "We like your recommendations. Would you like to come, would you be prepared to come and implement them?". I said, "Yes". And so, I took a leave of absence from the University of Oregon and went to Hong Kong for one year because I looked at this as a very interesting opportunity, not just to think about what you could do, but to do it'.

Mavericks and rebels

It was interesting to observe how non-conforming our leaders can be. At first glance, several leaders are non-cognate in terms of their formal education. So, Sir Michael Latham read History at Cambridge University and undertook a Diploma in Education in Oxford as part of his formal education. Nick Raynsford did a degree in History and Fine Art, whereas Kevin McCloud pursued a degree in the History of Art at Cambridge. Tom Bloxham studied politics at University. Although Guy Hazlehurst undertook a construction-related degree at the former Bristol Polytechnic (University of West of England), his motivation to complete the course as the top student was subconsciously driven by his desire to 'rebel' against those who supported him: 'I was doing one of the courses without A Levels and the fact that I ended up getting that prize just to prove people wrong'.

For Sandi Rhys Jones, gender stereotypes often inhibited her ability to engage in technical jobs. She recalled a summer placement for which she had applied at Boots the chemists: 'I applied for a summer holiday job working in Boots the chemist, had a successful interview, but the day I arrived to take up the job, I was directed upstairs. I said, "But upstairs are books, I'm working in dispensing and pharmacy". Apparently they had written to my headmistress for permission for my being able to work in the summer holidays and she said, "Oh, surely there's some mistake. This young woman is going to be reading literature at university, so she clearly should be in the books department". So, I spent my summer holiday in the books department at Boots even though my intention was to get experience in pharmacy and medicine, and I had fought to study sciences as well as languages at school'. Sandi went on to suggest how she was always slightly different from her peers at school, 'I read voraciously, I knew I could write well and I enjoyed languages. So the headmistress was right in that respect. I liked theatre too but I was also interested in knowing about the making of theatre sets and the lighting. So, although I could write plays and acted in plays and delivered plays, I was interested in the technology behind it'. And she complained how her brother 'always got the Scalectrix, Lego and Meccano. And I got a new frock! You know, it drove me dotty, it was very clearly defined. You know, pink things for a girl, blue for a boy. The boys got the toys and the girls got the pretty stuff'.

Paradoxically, for some of our leaders, their rebellion against the establishment has got them involved in the establishment itself. So, for instance, whereas Nick Raynsford spent his early career working for a non-governmental organisation (NGO) lobbying against governmental policy on social housing, he ended up in government himself. Amusingly, George Ferguson, past president of the Royal Institute of British Architects, always wears red trousers for official functions. George reflected on his unorthodox sense of dressing, 'The fact that I wear red trousers, it's not a political or artistic take. I think I got fed up with going to meetings where there were men in suits and I wanted to create an excuse for myself never having to wear a suit. I think there's a bit of rebel in me but nevertheless, I'm a rebel who tries who can be quite happy with joining the establishment'. Indeed, it is often much easier to instigate change from the inside. Chris Blythe showed his disdain for professional institutions in his younger days, 'I was just too much of a rebel. I didn't like the rigid discipline. You know, the hierarchy etc. you had to go on, you know, the senior partners and what have you'. Of course, he is now the CEO of the CIOB, although he prefers to be seen as a change agent in modernising the institute. Indeed, being embedded within the establishment does not necessarily imply conformity, as George Ferguson added, 'I always question authority. I think it is all questionable. I don't mean defy it for the sake of it, but I think we should always question authority'.

Personal passion for learning

What is striking in the interviews with all our leaders is the sense of passion for seeking improvements in the industry. We have already mentioned earlier about the personal passions of Jon Rouse and Nick Raynsford in social justice, which explain their endeavours in improving the housing situation in the UK. As Guy Hazlehurst puts it, 'I like to do work which somehow makes a difference after it is done. So, that's what I mean by impact. I suppose I'm not really that interested in making an impact in terms of personal profile, but maybe an impact that something you do, you have an influence upon something. Personally I thrive on that side, the ability to make a difference'. Passion stems from the Latin word *pasi*, which implies an element of suffering and perseverance. And for many of our leaders, there is undoubtedly a lot of perseverance (see Tom Bloxham's[48] top tips for success in Box 2.1), especially in relation to constant learning about the industry.

For example, Jon Rouse remarked, 'the only reason I would do something like that is that I'm genuinely driven by the fact that I think they are really fascinating. I think there's a lot of room for learning more within construction industry. It is a hugely misunderstood industry'. This need for learning influenced his decision to pursue an MBA in Finance. 'I decided to pursue for the love of academia by going to the University of Nottingham and doing a Finance MBA because I realised that actually my next step needed to be actually to manage a lead organisation. But if I really wanted to do delivery, then I had to do it with understanding of leadership and the big weakness was that I did not have the level of financial management skills that you would expect from someone who is leading an organisation so I chose a Finance MBA'.

By a similar token, the revelation of ignorance was also what drove Chris Luebkeman to go to graduate school, as he explained, 'my decision to go to Graduate School predicated on interviews in which I decided I did not know enough yet to consider myself ready to go to into practice. So, the kind of work which I perceived at that time I should be doing or

Box 2.1.   Tom Bloxham's top tips for success

- Whether you believe you can or cannot, you're right about it.
- Do not tell people, but empower people to achieve.
- Start with the end in mind.
- Start now and take risks.
- Hire only the best.
- Trend is your friend, but don't jump on the bandwagon.
- Strategise core skills better than anybody else.
- Turn employees into entrepreneurs.
- Keep it simple stupid.
- Look under every stone in your business and find the dirt.
- Make mistakes.
- Timing is everything.
- Be lucky, and grab every opportunity fate sends you with both hands.
- Persevere.

could be doing or would be doing, I did not feel that I knew enough. And that's why I went to Graduate School'. Kevin McCloud, on the other hand, was less strategic about his desire to learn: 'I tend to be quite reactive. I'm not a person who keeps a ten year plan. I am someone who just lets things sort of happen. I suppose I'm very spontaneous. "Oh, well, why not, that's interesting." My approach to my university life pervaded throughout my professional career as well'.

For Stef Stefanou, he values continuous learning, especially from others who he perceives as holding the beacon of exemplary practice. He commented, 'a company that had an impact on me was our biggest competitor for some time, O'Rourkes. Obviously, a lot of time, I used to see clients, and they would say to me, "Oh you know, I just saw the technical manager of O'Rourkes?", or there will be new developments that they are bringing out. I have to say, and I'm not embarrassed to say it, we created our technical department because of the technical department of O'Rourkes. We thought to counter it and to provide a service for the client that we did not think before. So, good competition also helps guard us. And now we have a technical department of five or six people'.

Another important observation relates to the modesty of our leaders in acknowledging their ignorance and their ability to tap into the knowledge of experts. Jon Rouse, for instance, considered himself to be fortunate to be working with great minds in the industry: 'I mean I was only 27 at the time and I just had great people to work with, people who really knew their stuff; people like David Taylor who was the Chief Executive at the time and David Shelton who was the Director of Development and people like Ralph Luke who is now in London and involved in the Olympics. These are serious players and they were hard-nosed professionals as well. David Taylor was an architect and the other two were surveyors and I didn't really have the same professional background. So, I actually had the opportunity to work with people who had been there and done it. Basically, I saw these guys and I realised there was a big gap between my background in the Civil Service, which was really about how to manipulate knowledge. And actually having knowledge, you know, actually being an expert, that is who these people were, they were experts and so, I decided to go back to college. And I went to night school, basically, and I combined the job at English Partnerships for three years doing a Masters whilst doing regeneration'.

Closing thoughts

The chapter began with a review of the salient points of leadership research, both in mainstream literature and construction management research. Although leadership scholars often glorify leaders on a higher pedestal, the reality is that most leaders have probably worked their way, even persevered, to the top. This book is really about empowering individuals who work in the construction industry to embrace the future (in a positive way). What this chapter has illustrated, hopefully, is that the 15 leaders often started from fairly humble beginnings. The fundamental difference, however, is that they not only have a passion to improve the immediate environment around them, but also have the wherewithal to follow this through. In a sense, the definition of leadership as

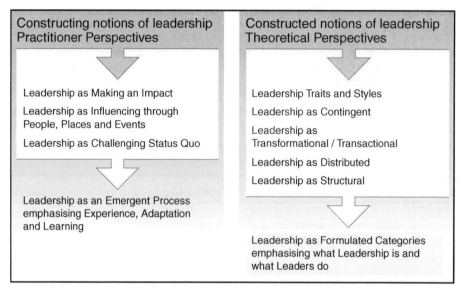

Figure 2.1 Comparing practitioner and theoretical perspectives of leadership in construction.

someone who influences holds true. That said, it is perhaps more crucial that leaders want to influence or create an 'impact' to use Guy Hazlehurst's terms. In this chapter, we have also refrained from using the interviews to validate the theory of leadership. Instead, we have consciously sought to set our participants' perspectives in the context of their career and life journeys to get an intimate understanding of their leadership trajectories. Clearly, aspects of all five clusters of leadership theory do feature in our interviewees' lives. It is central that we move away from understanding constructed notions of leadership as a thing, to constructing a more fluid idea of leadership as an emergent process. We have illustrated this in Figure 2.1.

The extant academic literature on construction leadership has hitherto been somewhat prescriptive and perhaps over-intellectualised. Much research effort into explaining leadership has tended to follow a path of identifying key traits, at a given point in time. Of course, we are potentially culprits of this in our analysis of leadership through the lens of our influential participants. Nonetheless, it is our intention to simplify, yet elaborate on, and maintain the dynamic nature of the workings of leadership through the stories of how our leaders have been shaped by their forbearers, and how they then continue to influence others in their professional careers. The review revealed a need for more in-depth analysis into the development of leaders by examining their life histories to establish a broader social view of how they have developed as leaders of the industry. After all, people, places and events matter over time as our analyses have shown. Our 'leaders' are certainly well connected in the industry, and, as we have found, frequently network with one another. As Bob White neatly summed up, 'At the end of the day, in business and everything else we do, it is about relationships. It's not about materials and structures. It's about how people can best work together and make things happen'.

Knowing people is only one half of the equation. Our leaders are active participants of the industry who constantly challenge the status quo. Indeed, our leaders come across as perceptive individuals who adapt to the multitude of situations in which they find themselves. More importantly, they seize every opportunity with both hands whenever opportunity knocks, and they know how to tap into the knowledge of the people with whom they connect. In fact, our leaders have demonstrated that they not only influence the context of their practice, but are also comfortable in allowing events to shape their thinking. We will discuss the interplay between people and systems[49] as we move into the later chapters on the sustainable development agenda and governance.

To some extent, this chapter has exposed the value systems of our leaders. Interestingly, all of them enjoyed being asked questions about their pasts. This is probably one of the rare occasions in which they take some time out of their busy schedules to reflect on their personal development. Perhaps, this reflexivity is needed for anyone wishing to cast a line into their own futures. However, things must move on. In the next part, we will leave the past lives of our leaders and delve deeper into their thoughts about current issues and future challenges.

Part 2

Eliciting the future

Chapter 3

Developing a sustainable future: theoretical and practical insights into sustainable development

'In a moving world readaptation is the price of longevity.'

George Santayana, 1863–1952

Chapter summary

Any discussion about the future invariably evokes thinking about the issue of sustainability. So it is unsurprising to find that our leading figures have placed much emphasis on sustainability issues that impact on the longevity of the industry. At its core, all our interviewees recognise the importance of people in setting any debate about, and interventions on, securing a sustainable future. There is the acknowledgement that physical structures of the built environment are meaningless if not for the people who design, construct and use such facilities. In shaping the future of the industry, it is therefore critical to consider how the industry contributes to the livelihoods of people living and working in communities.

The concept of 'sustainability', nonetheless, is loaded with much complexity, and in turn is fraught with tensions and contradictions. A critical paradox is the dominance of the economic perspective that underscores the theory and practice of sustainability; where the economic imperative drives much progress made in understanding the agenda, its narrowly rational approach often impedes real action in the quest for a sustainable future. Admittedly, such tensions and contradictions arise because of the difficulties in framing an understanding of the often-uncertain benefits to future generations and the pressing need to satisfy demand in the present time. Consequently, the focus is often misplaced, away from the real need for sustainable development, and instead emphasises measurement of an agenda that promulgates the monetary valuation of a set of arbitrary terms, with relatively less emphasis on the less tangible, more experiential aspects of human well being. So ironically, although both our interviewees and the literature acknowledge the importance of the social dimension, the prevalence of the economic viewpoint means that the social remains largely elusive.

Constructing Futures: Industry Leaders and Futures Thinking in Construction   Paul Chan and Rachel Cooper
© 2011 Paul Chan and Rachel Cooper

In this chapter, we first present an analysis of our leading figures' views about sustainability. Here, we observe that our leaders meld together economic, social and environmental concerns – i.e. the triple-bottom-line – when thinking about industrial development over a sustainable future. Amidst the conversations lie their thoughts about the role of governments, industry and the education sector in ensuring that a holistic approach can be garnered to meet future challenges of sustained competitiveness, skills capacity of the workforce, climate change and environmental preservation, and the building of sustainable communities. Whereas the need for such a holistic, joined-up approach is acknowledged, the way in which our interviewees frame their thoughts seems, at times, to pragmatically focus on what they individually can achieve, often driven by their pet passions and restricted by constraints of reality.

An attempt is also made to compare and contrast the thoughts of our interviewees with a review of theoretical perspectives on the subject of sustainability. Drawing on the conceptual framework provided by the late Professor David Pearce,[1] which includes man-made, social, human and natural capital perspectives of sustainable development, it is useful to observe that there are many overlaps between the theoretical and practical insights of sustainability. Fundamentally, the core argument in this chapter is that knowledge about what sustainable development really means remains incomplete. What is important is not necessarily the definition and measurement of aspects of sustainable development, but that it is critical to examine actions that people take in affording a sustainable future. Furthermore, the various capital perspectives of sustainable development are highly complex and interconnected, and tensions and contradictions mean that trade-offs between various perspectives are inevitable. Finding a universal panacea for sustainability is therefore inappropriate and impossible. Instead, it is more fruitful to frame the agenda as a collective effort towards building sustainable futures and this requires an institutionally coordinated response to engage the state with businesses and communities. The chapter also urges greater research efforts to understand sustainable development as an emergent process.

The key issues discussed in this chapter are as follows:

- Economic considerations often dominate discussions about sustainable development. This is paradoxical. On the one hand, economics provide a lingua franca for politicians, business and community leaders to discuss impacts of, and interventions on, meeting the sustainable development agenda; yet, the emphasis on monetary valuation prevents genuine progress made in securing benefits of a sustainable future.
- The economic perspective results in the obsession with measurement with relatively less emphasis on what these measurements mean for the future well being of people and how these materially contribute to effective policy formulation and implementation.
- Sustainable development is a complex concept encompassing a number of interconnected facets. Understanding trade-offs in decision-making is therefore critical, especially where there is a fragmented landscape of stakeholders involved. There is a need to consider the socio-political and economic structures of decision-making, and opportunities for joined-up thinking and action need to be explored.
- The social dimension needs to be brought more to the fore. Sociological and psychological disciplinary knowledge can be mobilised to better understand the nature of

human agency and behavioural change in delivering the sustainable development agenda.

Introduction

If official statistics are anything to go by, the British economy has seen, up until the current financial crisis, unprecedented periods of economic stability since the early 1990s. At first glance, this should be welcomed by an industry that is often used as an economic barometer for governments around the world.[2] However, 'the fortune of firms in the industry is hostage to indicators such as interest rates, unemployment, inflation and economic growth. All these factors have been encouraging for the industry in the early 2000s but questions remain over whether these conditions are sustainable in the long term'.[3] Certainty about the global economy is indeed questionable at the present moment. The world has experienced one of the deepest recessions since the Great Depression of the 1930s. In the UK, the Monetary Policy Committee (MPC) of the Bank of England has had to maintain interest rates at its lowest rate of half a per cent since the beginning of 2009. There is also political uncertainty looming, which does not bode well for predicting the future of public spending. Following an intense period of fiscal spending to quell the tide of economic decline at the onset of the global financial crisis, cuts in public sector spending seems inevitable. Possible restraints to future investments on key areas of healthcare, education and housing are likely to threaten construction activity. This is likely to prevail for the foreseeable future in many countries across the developed world. Jobs continue to be lost in the manufacturing industry to low cost-base countries, and the global economic recession has seen the onslaught of jobs across the financial and services sectors and a dismantling of employment and social security. These surely make for uncomfortable reading about a tightening global economic situation.

On the other hand, there is still hope for the construction industry with high-profile projects such as the Olympics in 2012, its associated decommissioning work after the games and potential for reinvigorating investment activity in the private sector. The swearing in of the Obama administration in the USA has opened up opportunities in terms of international relations, and the US renewed commitment to the environment potentially translates into jobs in the green economy. The key challenge, of course, is its sustainability. It is the challenge of creating a sustainable future that has driven the publication of the seminal Pearce report.[4] Furthermore, researchers have been mobilised, through initiatives like the Engineering and Physical Sciences Research Council (EPSRC) *Big ideas* programme (www.thebigideas.org.uk), to examine sustained competitiveness of the construction industry. Given the gravity of the issue, this chapter is therefore devoted to discussing our leaders' thoughts on the issue of sustainability, an issue that unsurprisingly featured prominently in our interviews.

In this chapter, we first present an analysis of our interviewees' perspectives on 'sustainability', focusing on the critical issues emerging from the interview data. These include the significance of interactions between people and places, the role of the government and industry in responding to the growing agenda of sustainable development and climate change, and the need for shifting thinking in education and research. For our interviewees, despite coming across as being passionate about what the future could mean for people in communities, many also conceded that the economic imperative ultimately determined the extent to which the sustainable development agenda was embraced and delivered. There was general acknowledgement of the harsh realities of needing to survive in a competitive marketplace, which invariably drives corporations to focus on the profit-making motive. Consequently, the potential for the industry to contribute to the creation of a more socially and environmentally just world can, at times, be thwarted. At first glance, it would seem that our leading thinkers have merely reiterated the rhetoric that sustainable development matters, and accepted the dominance of the economic perspective that perhaps limits progress made on this agenda. It was also notable that relatively less emphasis was placed on environmental concerns; nonetheless, this is possibly attributed to a lack of comprehension of the evidence and scientific knowledge that is advancing in this area.

So, in the latter half of this chapter, we review the state of scientific knowledge around the core pillars of sustainable development by tracing four capital dimensions of sustainable development according to Pearce.[5] These four dimensions – physical, social, human and natural capital perspectives – are contrasted with our leading figures' understanding of the triple-bottom-line approach towards sustainable development. On reflection, the responses by our interviewees are mostly pragmatic, and whereas there is recognition of the various dimensions of sustainable development, there appears to be a chasm between the longer term theoretical and policy aspirations of sustainable development, and relatively shorter term practical considerations and actions at the grassroots level. Although a critical review of the scientific literature reveals a set of rather complex and interrelated theoretical concepts, our leading thinkers seem to be very much focused on what actions might appear feasible so that these can be enacted in the present to move things forward. Yet, the danger, of course, is that any action based on partial and simplistic treatment of what is incomplete knowledge in the scientific field would result in a lack of holism in tackling the problem of 'sustainability', and, at worst, lead to detrimental effects. Such partiality was observed in the analysis of the interviews, as each interviewee tended to approach the agenda from their pet passions, emanating in part from their personal and career histories examined in Chapter 2. Consequently, our leading thinkers appeared to frame their understanding of sustainable development in terms of single contemporary issues – such as reduction of carbon emissions, community development, skills development, and people and diversity management – as opposed to deeply exploring the connections and intersections across these issues from an academic, theoretical perspective. Admittedly, our interviewees considered the industry to be limited in terms of how it can holistically make sustainable development agenda materialise, as they suggested that more needed to be done in relation to building up knowledge in this area through basic research and joining up efforts in policy-making at government level.

Connecting people, profits and planet: the rise of the sustainability agenda

As outlined previously, the principal purpose of this book is to provide a personalised view of the future of construction through the eyes of leading influential figures in the UK.

Many foresight reports present snapshots of future scenarios, yet these are often divorced from any adequate explanation of how these scenarios have been created or from what value systems these were derived. From the interviews, it became apparent fairly early on that, for our leading figures, any thoughts about the future were mostly about securing the perpetuity of human existence in this world. Of course, we have come to frame this in the notion of 'sustainable development'; and although the term 'sustainability' was not always used per se by our leading figures, analysis of the interviews indicate that this is a significant cross-cutting theme that is of concern to them, and so this will be elaborated here. The concept of sustainable development has gained much scholarly attention since its genesis in the late 1970s, and it is interesting to note that there is a great deal of congruence between the theoretical framework of the triple-bottom-line (i.e. economic, social and environmental drivers) and what our leading figures thought about 'sustainability'. Virtually all our leading thinkers also considered the future in the context of people, planet and profits.

In the next few sections, we make sense of our leading thinkers' perspectives on sustainable development, and re-present their thoughts along three key themes. We note initially the credence placed on the importance of people and places, the role that the government plays in developing policies and investing in infrastructure, and the actions the industry can take to address the sustainable development agenda. We also take a brief look at the role of education and research in sustaining the future capacity of the sector as we present the need for more joined-up thinking.

Interactions between people and places

One of the most striking observations made in the interviewing process was how every interviewee seemed to put people at the heart of any conversation made about what they do in construction respectively. In this section, we outline their views regarding the interconnectedness between people and the built environment through a number of emerging issues. These critical issues point to the importance of building communities, as buildings as physical structures alone are meaningless without the people who design, construct, occupy and use them. The discussion also considers the need to balance both economic and social aspects, as these aspects contribute to the success of any community in the wealth they create by giving people something to do with their time. The need to search for local solutions to local problems is also emphasised, and the section talks about how crucial it is to energise local activism and mobilise networks of influential people in the quest to create sustainable communities.

It's all about people, stupid! The essence of communities

Indeed, the output that the construction industry produces, according to Kevin McCloud, is not just about buildings, but 'the relationship between human beings and buildings' as he argued that this really is the central principle of managing any design and construction of the built environment. In fact, Kevin McCloud asserted that the popularity of his *Grand Designs* programme lay not in the designs themselves, but the fact that the centrepiece of each programme revolves around the lives of the people that design, construct and live in these buildings: 'In as much as 1% of people that have seen our programmes work in the built environment, another 3% are really interested in the architecture and in the design and the rest of the 96%. . .well, they are just coming along for the ride. They love the story-telling. They like a good yarn, but in so doing, what they are doing is, I believe they are learning about buildings'.

The significance of the social aspects of construction can never be downplayed. The built environment cannot simply exist as stand-alone physical objects, without consideration of its use by the people living in the community. Throughout the interviews, it is clear that the notion of the community is of particular concern to our leading figures, as all our interviewees considered how the lives of people are intertwined with the physical objects created by the industry in what is essentially a community. Kevin McCloud, for instance, cited French colourist, Jean-Philippe Lenclos to illustrate how buildings and people are interwoven together: 'If you wanted to choose a colour for your front door, get your neighbour to choose it'. Tom Bloxham was clear that his work as a property developer was not just about the creation of another building, but to create 'a whole new community, a whole new village, a whole new town'. He was especially inspired by Saltaire (elaborated later in this chapter), and romanticised about how people in medieval times got it right about what sustainable communities really meant: 'Any medieval village is mixed use. You've got the baker and the shops downstairs, with people living upstairs. And if you go to any European City today, I mean, every building is mixed use'. For Tom, mixed use development is a sensible way to bind the economic and social aspects together; after all, people need to find things to do to occupy their time, whether this is through employment or in the places they live. He was, however, rather uncomfortable with using the now popular phrase 'sustainable communities', as this was somewhat politically loaded jargon. Tom stated, 'I'm not even sure I like the phrase "sustainable communities". For me, I just build places'. And a good place, according to him, requires a few key ingredients, including 'a good modern design, mixed uses, innovative contractual forms, and a cheap stock of old building that no one knows what to do with and of course, other people's imagination'. The challenge, as Tom sees it, is how city centres can be rebuilt and revitalised from time to time, and how towns and cities are integrated.

Sustaining a community by balancing the economic and social aspects

Other interviewees also echoed Tom's view that an important yardstick for measuring what a sustainable community does is simply whether people living in these communities have enough activities to occupy their time. And, of course, economics play a significant role here. For example, Jon Rouse waxed lyrical about Bradford, an industrial city in West Yorkshire in which he spent a lot of his formative years. Tracing the downfall of Bradford and the rise of urban deprivation in what was once a wealthy city fuelled by the woollen industry, Jon recounted, 'Bradford was the most beautiful city in the country. It was a Victorian city. Okay, you can go to Bradford today and you wouldn't believe that, but if you look at the old photographs of Bradford up to 1950s, it was a very wealthy city. The woollen industry made it a very wealthy city and the architecture reflected that. And you wouldn't believe it today, but Bradford used to be the best destination choice in West Yorkshire. It wasn't Leeds, it was Bradford probably until the 1960s. Then unfortunately, the collapse of the woollen industry and the very poor infrastructure that Bradford's got led to its economic collapse'. So, securing an economic future seems to be a critical first step in creating a sustainable community, as according to Jon Rouse, the rest (e.g. leisure activities, maintenance of fine architecture and infrastructure development) will follow.

Alan Ritchie from UCATT told a similar story of the thread mills in Scotland, focusing this time on the working conditions of the workforce: 'If you want to go back in history to the old thread mills, right? James was the owner. He owned mills up in Scotland. So, what he done is, at the time there were children working in this industry and he built these big thread mills up in Lanarkshire, Scotland. And Robert Owen, he built the schools for them. He gave them free education. He built homes for them, you know, proper facilities in them and his production went through the roof. And so, what I'm saying is, this shows that how you treat your workforce, you get a better response'. Of course, as we shall see in the case of Sir Titus Salt in Saltaire below, this is a prime example of British philanthropy in centuries gone by. And although economics matter so that people in communities feel secure with job prospects to expend their productive capacity, it is also about ensuring that people are happy with the living conditions with which they have to contend. What Alan Ritchie emphasises are just these conditions, through the local service provision in the communities in which people live. It is perhaps worth noting that combining both economic and social aspects to create sustainable communities is easier said than done. As this chapter unfolds, it will become clear that both these perspectives are often treated at extreme ends in policy debates, and, as these develop, urban deprivation still exists, especially in communities across the UK where the main economic activity has declined over the years (e.g. in the cotton, mining and, more recently, traditional manufacturing industries). Raising aspirations, therefore, continue to be a sticky challenge with which policy-makers grapple. It is interesting to observe, nonetheless, that our leading figures often reflect on stories in the past and appear to credit these as the 'good old days.' Perhaps moving on into the future is a much more challenging, incremental thing to do.

The significance of the comparative: knowing where the baseline is

One of the critical issues confronted by policy-makers and professionals in the industry is knowledge about whether the communities we create are necessarily successful. This is problematic, as it deals with knowing what the comparative is. Yet, if one does not have anything to benchmark against, it can be difficult to recognise the opportunities that are available elsewhere. Jon Rouse's life story is interesting in this respect. Jon, being exposed to urban deprivation at an early age, was, of course, attracted to a lifelong career in facilitating urban renewal: first at CABE, then at the Housing Corporation, and, most recently, as Chief Executive of the London Borough of Croydon. However, deprivation needed to be understood in relative terms according to Jon: 'I think from an early age, I wasn't conscious of deprivation at the time. You know, although Bradford and Barnsley were poor places; people did not have a lot of money. But because when I was living in Barnsley, both of my parents were working. My dad was a social worker, he wasn't earning a huge amount of money but we were actually one of the wealthiest families in the area. However, when I was eleven, my parents – mainly for the sake of mine, and my brother's and sister's education – decided to move away from Barnsley and we moved to leafy, suburban Northamptonshire. When we moved to Northamptonshire, we nearly became the poorest family'. It was in Northamptonshire that Jon understood what the North–South divide meant in terms of the wealth and income gap, but, more importantly, he experienced firsthand the impacts of class division and social exclusion. He described himself as the 'Northern kid moving into a very leafy suburban contrast', and explained that the contrast was what motivated him to tackle such social division in his career: 'I was in my early teens, and I also gained some friends at that time and there was a couple of friends who had also moved down from the North to the South and we were all pretty intelligent kids. We were all from the North, whose families had moved down and we became very close friends and we were very precocious and we would comment and debate from a distance on the things that Thatcher was introducing in 1979'. Therefore, it is having personal knowledge and experience of the comparative between rich and poor that arguably equipped Jon Rouse to tackle poor living conditions in his long-standing career in creating sustainable communities.

The significance of the comparative also featured in Tom Bloxham's interview, as he traced the development of another Northern city, Manchester. He explained, 'The key moment in Manchester wasn't the Commonwealth Games [in 2002], it wasn't the IRA bomb [in 1996], it was losing the bid for the Olympic Games [in 1993]. We only lost the Olympic Games, but we wanted to celebrate it because for the first time, the people in Manchester realised that we weren't competing with Barnsley or Bradford or Stockport. We were competing with Los Angeles and Sydney and Barcelona'. It would seem that having these comparatives mattered in raising the aspirations of Manchester, and it was precisely this lift, Tom argued, that allowed Manchester to submit a winning bid to host the Commonwealth games in 2002. This, together with a sympathetic planning system that is willing to engage with the community it serves, add to the critical success factors of how the agenda of sustainable communities and the tackling of sticky challenges, such as urban deprivation, can be met. In the next subsection, we turn our attention to how the creation of sustainable communities can be facilitated by mobilising networks of influential people.

Creating sustainable communities by energising interactions with people

Energising local activism and the mobilisation of networks of influential people are also critical in getting schemes off the ground. Describing how he got involved in regeneration work in Liverpool, Tom Bloxham elucidated, 'In Liverpool, a guy who worked at the Planning Office, called Bill Maynard – he actually now works for us – knocked on the door and said that although Liverpool had this plan of making a Quarter, nothing has actually happened and he was looking around for help. So I went out and he talked me into doing something and getting stuck in. They came to see me to see what it was and gave me a grant of ten grand or something to do it. And he said, "Here was somebody actually doing something, you know, not just talking about it. You know, spend loads and loads of money on studies, on architects". And that was how it happened. And then, he started getting involved and the relationship developed and he then turned round and said, "How else can we help you? You're doing great with that quality that you're doing and you know, that was useful and so on". And in Manchester, I think the process was similar, i.e. to get a building up, applied for a grant, go through the planning process, talking to people, making a difference to everybody there, etc. You know, it's important to make it a business to get to know people, to get to know them personally'. Although we deliberately steer away from offering prescriptive methods in this book, Tom's explanation suggests once again that the success in creating a sustainable community is very much dependent on the effective mobilisation of people, especially of a passionate team of people wanting to do something to give back to the community.

Energising such local activism sometimes also meant that one needed to become a person of influence, and not just rely on the influence of others. George Ferguson reflected on his early days in political activism, which began in the humble roots of a campaign to preserve the heritage of the built environment in Bristol. George explained, 'I have a slight aversion to party politics. In 1979, if you remember, Thatcher was elected and, between 1979 and 1983, I was concentrating on my architectural practice and the building aspects. But by the time it came to the first election after Thatcher was elected, i.e. in 1984, I was so spitting mad with what she was doing and with what her Minister for the Environment, "Riff-Raff" Ridley was doing in terms of encouraging out of town shopping. I could see him wrecking our city. I could see them absolutely wrecking the city. I mean, Heseltine was to come in later and repair the situation with initiatives in Liverpool and elsewhere. So, I agreed to stand for Bristol West for Parliament against the Conservative, William Holdsworth, who I quite like actually. So, it wasn't a personal thing, but I was absolutely outraged and I had become a father by then – I got married in 1969, and so I had kids – and I was, you know, thinking about their future. So, I agreed to stand for Parliament'.

Although he did not get elected then, that episode in George's life initiated him into a lifelong expedition of personal activism to campaign for the preservation of the English heritage. In his helm as the President of the RIBA, he continued influencing policy-makers to take the idea of sustainable communities more seriously: 'I felt the RIBA itself was not paying enough attention to the preservation of historic buildings and conservation. It did not pay enough attention to planning and urbanisation and so, those were the two things I really concentrated on, particularly urbanism. And, you know, I got on extremely well working in parallel with government ministers who were trying to move towards sustainable communities'. George too found the term 'sustainable communities' politically loaded, and often resented the fact that such slogans were merely deployed as a rebranding exercise of what had gone on before. He lamented, 'Now, I have criticised them saying that really the phrase of these sustainable urban communities were essentially to build housing estates'.

Making sustainable communities successful: tensions between the global and local in gaining consensus

What makes a place successful? The assessment of success has been known to be fraught with problems. Notwithstanding the ability to benchmark against comparatives, and in spite of the presence of local activism, tastes and perspectives as to what is good for people and places are variable. Gaining consensus is never straightforward. However, George Ferguson takes a pragmatic approach here. 'You can get some really banal buildings that people will say, "Oh, well, the new houses are a horrible place". If you do a survey, you'll find the majority of people will agree on what's beautiful and what's not.

There will be some disagreement somewhere in the middle and of course, there are some extreme examples like the very high profile projects e.g. Lloyds of London building by Richard Rogers was one quoted in two different polls, both the least favourite and the most favourite contemporary building. It got both. It won both. So, there's that extent we've got different tastes but I think, generally, we can get a much higher level of agreement in terms of what people consider beneficial.' For George, instead of emphasising differences, it is more often than not easier to identify on what most people would agree. However, George, along with many of our interviewees, reiterated the importance of working out localised solutions for local problems when searching for consensus in the creation of sustainable communities, 'What we have to be aware of though, I think, is making everything the same. I don't think that something that is beautiful in Singapore would be deemed necessarily beautiful in Cornwall'.

Bob White also suggested that the pursuit of the sustainable communities' agenda needs to be tailor-made to the perceived problems that need resolving, as he explains, 'it would depend on the type of projects; there are certain types of projects which invariably are just one-offs, others that are not. And so this determines whether we use the national or maybe even an international supply chain. If you're building the Scottish Parliament Building again, and get it right this time, you could use Mace or you could use the French company Bouygues, or Balfour Beatty or whoever. If, however, you are constructing 100,000 houses in the Thames Gateway, it is very likely to be a localised participating activity. So, you create a different environment and a different structure for the type of operations you've got. Certainly, housing and schools should be a local product, by and large, because this is what would employ the local community and create the local community supply chain. By the way, you then teach them how to extend their activities, improve their skills into extensions, repairs, maintenance and all that sort of stuff. So, you actually create a holistic approach'. Therefore, what defines success of a sustainable community is contingent on the scale of the activity and how the inherent tension of seeking localised solutions within a globalised view can be resolved. Yet again, we see that people lie at the heart of how one defines what is 'sustainable'.

Continuing on the theme of the global–local paradox, and reinforcing the significance of human relations in the sustainable development agenda, Guy Hazlehurst focuses on the need to encourage localised solutions in the development of what he termed as a 'sustainable skills landscape'. Guy argued that global approaches to solving localised problems may be deficient after all, and suggested that a case-by-case specific approach to tackling problems about sustainable development is critical: 'There is of course the danger of sensationalism of talking globally about skills shortages and skills gaps. Where are we going to find the skills to meet these shortages and gaps? You know, where are the hotspots? And the hotspots are either going to be by region or type. It's hyped up, anyway. And you know, as you say, in terms of sustainable skills, we should be bothered with it, but we don't know what they are yet. So, if you're building the armadillo-shaped concert hall like the Sage in Newcastle-Gateshead, it would matter where the people come from, and where they go because if you are going to create some legacy skills, then those legacy skills will be following all the labour around and maybe what we should do is look at the market, look at the work profile and generate the sustainable jobs and focus on those. So, that's creating sustainable skills. The people who train where they are working in the same area, and living in the same area, export those skills first of all, locally and then look outside'. So, one can again see the constant connection made between the economic and social aspects; the creation of jobs is dependent on the availability of skills to deliver specific projects, and its longevity is dependent on whether skills development opportunities are accounted for or not.

Guy also noted that a lot of construction labour is local anyway, contradicting the view that the construction workforce is often mobile and transient. Drawing on the migration study undertaken by ConstructionSkills, he observed, 'I think the issue of mobility is one that we are only just starting to grasp because there is this fear that the construction workforce is incredibly mobile, in fact that is not the case. A great proportion of the construction labour force is actually, pretty much indigenous. It likes where it lives. You know, it doesn't tend to move'. Consequently, this lends further support to the need to frame localised solutions in the creation of sustainable communities.

Guy was, nevertheless, cautious about only advocating the local economy, as he acknowledges that capital, goods and services can be very mobile in the globalised world we live in today. 'Take glass cladding for example. I did a project in Bristol which showed that the nearest firm where you could get a particular type of system to work was not from Bristol but Southampton. Now that company is not going to employ, generally, Bristol people. They'll bring them from Southampton. They'll come to the project site in the morning and they'll leave in the evening. And yet if you look up the Bristol phone book, there are Bristol cladding companies. The thing is it's not got the right type of companies who can hang that kind of system on that kind of building.' Guy added, 'There's also a dynamic that is often missed in construction which is that, and I think it's getting people to understand [. . .] it always amazes me when John Prescott is apathetic on the material of the Dome coming from Germany. But there are only so many places that you can get Teflon-coated plastic in the world and if you can't get it in the UK, then why worry that it comes from Germany because there's not that many places that can make it. So, there are things that we can and should worry about. As I say, if a local ducting firm creates local jobs, then we shouldn't perhaps worry about that and sort of focus our attention on other areas'.

To summarise, human relations matter significantly when it comes to shaping the future of the built environment. It is critical to consider how the design and construction of buildings contribute to the livelihoods of those who work and live in communities. Yet, how communities interact with the built environment can be somewhat complex; it is not simply a case of stating the importance of people. This is obvious! There are trade-offs that need to be accounted for in the pursuit of sustainable communities, including balancing the need to generate wealth through job creation and enabling people to do something meaningful to occupy their time. It is also about balancing the global view with a need to provide local solutions to local problems that meet the local requirements of the community. It is about ensuring that the sustainability of job prospects is reflected in the skills that are either available or have the potential to be developed in the local communities, and it is about how communities can be mobilised to take an interest in co-creating the future.

Left to its own devices, however, it seems that the economic imperative still dominates the thinking process of the professionals that we interviewed. The corporate decision-making process is often driven by the profit-making motive, which impacts on the nature by which labour, goods and services are procured. Linking this economic imperative with the social aspects of the sustainable communities' agenda might require a more holistic, institutional response, which we shall elaborate in greater depth in the next chapter. Although people do apparently matter in our interviewees' perspective of what sustainable development in general, and sustainable communities in particular, are about, it is still the ability to generate jobs and keep the order books filling up that can maintain the aspirations of communities and quell the tide of deprivation. However, communities run the risk of simply looking back to past glories and pursuing a 'no-outsider', parochial mentality that encourages insularity. There is often a need to ensure that a global, longer term view can be ensued. In the next section, we consider the role that political leaders can play in connecting up communities through infrastructure development and how governments can help (or even hinder) in this respect.

Role of political leaders and infrastructure development

'The heart of the industrial revolution was based along the canal corridor going from Liverpool through to Hull, connecting the Irish Sea to the North Sea. And so, the economic collapse was experienced by all those cities in different ways. Manchester was cotton. Bradford was wool. Leeds was just the worst at industry and so on. And Liverpool and Hull with the ports at either end. And you know, the ones that recovered quickest were basically Leeds and Manchester because they had the best North/South links, primarily rail as well as road. Yes, but they both had links to the M6, the M1 [motorways] and the airports and that's really why.' Previously, our interviewees have established that to sustain a thriving community, there needs to be continuous investment and renewal of infrastructure to facilitate economic progress. Here, Jon Rouse reiterates the importance of infrastructure development by tracing how this can help a community weather economic decline; this remains equally relevant today as governments across the world are attempting to revive the global economy by investment in infrastructure development in some shape or form. In this section, we outline our interviewees' views on the role that government plays in terms of investing in the built environment and regulating construction activity to safeguard well being for all, whether this is manifest in the way governments act as effective guardians of regulatory frameworks for encouraging behavioural change, or the way governments are open about their policy intent, or the way governments can reap a deeper understanding of human agency so that policy instruments can be designed to incentivise sustainable behaviour. These will be elaborated in turn within this section.

The ticking of the energy time-bomb: the problems of technological advances and depletion of energy sources

Although the theme of climate change did not feature too prominently in our interviews, concerns were raised about energy use in the sector alongside the fear of the future of energy sources, especially in relation to how this is connected with infrastructure development. George Ferguson, for instance, remains anxious at the rate the built environment is consuming energy in terms of its development and its use, especially in developing parts of the world: 'I'm afraid what I see in China now absolutely horrifies me. It's not making a good place. There's uninsulated building; they are burning up a ridiculous amount of energy'. He added that one should avoid a 'Because we can, we shall' attitude, particularly in relation to the use of new technology and materials: 'I think we have developed beautiful places that are distinctive. Of course we haven't been able to, in the past, move stone and other materials right across the world, so they used local materials, local details and people who could work with local weather conditions etc. Now, I am not arguing against the employment of new technology, that we shouldn't exploit new technology. I'm just arguing against the extreme way with which we seem to be adopting new materials and technology now'. So, for George Ferguson, technological advances are not inherently unproblematic, as greater efficiencies accrued in one area might result in the creation of problems elsewhere. What George argues for is a reflective perspective on how technology might lead to improvements in overall well being, and the question on energy consumption remains a pertinent one.

Yet, technological advancements can also be a solution to the crisis about energy. Certainly, Guy Hazlehurst was gravely concerned about the reliance on other countries for traditional energy sources, and, in fact, suggested that this reliance might just bring about an economic downturn after years of relative stability: 'I think that there is the chance of some sort of external shock to come, like Ukraine turning off the gas supply. You know, that probably is indication that the economy can no longer be sustained, and suggests a certain level of susceptibility'. Interestingly, Guy did allude to the prospect of an economic recession during his interview in early 2006 brought about by an energy crisis. This was, of course, at a time when the world was seeing a surge in world oil prices, which probably influenced Guy's thoughts about the implications on the economy. He also expressed anxiety on the levels of public sector investment: 'We may have over-done our investment in the public sector in a way that may not be sustainable and you know, recession prospect scares you', but thought that any recessionary trend will not see the 'same sort of calamitous pressure [as] in the last recession. I think firms have readjusted; if you look at the last recession, it really was a recession in the commercial sector and a combination of external factors. External shocks aside, I think a lot more stable, I think the chance of us having a 1980s, 1990s type collapse is less likely'. Indeed, with the benefit of hindsight, Guy was certainly accurate in suggesting that external influences are increasingly critical to the sustenance of national and local economic regimes, but what he could not have appreciated then was how rapidly coordinated and widespread the financial crisis has transpired in recent times, in part due to imprudent commercial decisions yet again. We now turn to the role of governments in regulating commercial behaviour, and question the extent to which governments can do this effectively given the law of unintended consequences and the implications of globalisation.

Can governments act as an effective guardian of regulatory control?

The role that government plays is undoubtedly a significant one, as it is well known that governments are a major procurer of construction goods and services. In the global financial crisis towards the end of the noughties, governments across the world are again mustering their power to invest in infrastructure development in an attempt to kick-start the economy again. Furthermore, governments still retain their power in the legislature and the regulatory frameworks that govern construction, which serves as a potentially notable influence on behaviour in the sector (see Chapter 4). As Alan Ritchie insisted, 'What I will say is that the biggest client in the construction industry is the Government and in their procurement policy, we should be determined to set the standards in which companies will tender for'. So, as a major client of the sector and in its enforcement capacity as regulator, the government is instrumental in ensuring that working practices can result in sustainable development through its provision of infrastructure and the built environment.

However, the role of the government has come under criticism by a number of our leading figures. First, there is a feeling that political agendas sometimes get in the way of genuine progress made in sustainable development. The complexities of government organisations coupled with the turnover of personal interests and portfolios can mean that the construction industry does not get the attention it deserves. Alan Ritchie commented on his experience when discussing contractual arrangements in the Scottish Parliament project, '[UCATT] approached clients in some of the contracts and we raised a question with the Government in the Scottish Parliament about how bogus self-employment was going to be addressed. The minister gave a reply saying they cannot do anything because of the European Competition Directive, which was nonsense. So, we got the lawyers on to them. The document was that thick, we read through it, and there was nothing about the Competition Directive in there. And it was like a "Yes Minister" answer: It came from a Civil Servant who knows nothing about the industry and was looking for a cop out'. The issue of self-employment raised by Alan Ritchie underscores an important area for sustainable development, as it can adversely affect both employment relations and the extent to which skills development takes place for the sustained capacity of the workforce, an issue that can only become ever more critical, if under threat, during an economic recession. The UK government's neo-liberalist, arm's length approach has certainly resulted in deregulation of the labour market over the years, which, in turn, led to the disappearance of direct labour employment in local government and the rise of what Alan Ritchie called 'bogus self-employment'.

Sir Michael Latham outlined this development in greater detail: 'It's changed to some extent. Thirty years ago, when I first joined the Building Trade Federation in 1967 and used to go to Council meetings, we would have a considerable number of medium-sized and small firms on the Council who employed quite a lot of direct building labour. There was less sub-contracting than there is now. There was particularly more direct employment. That began to change, basically, in the late 1950s, early 1960s. What gave it a big heave at the time was in 1966, the Labour Government introduced the selective employment tax, which was a charge on employees and it was subsequently abolished by Heath's Government in 1971. As it was, the selective employment tax sent a message to main contractors, who, as they were then called, the main contractors, that they would do better if they didn't employ people directly. It was a long precursor of [the Construction Industry Scheme, CIS 714] and in fact, what then happened were two things. On the one hand, the employers had a financial incentive not to employ people if they could avoid it. Apart from any other consideration, if they suddenly found they had run out of work, they had to tax them and pay the redundancy pay and so that would come in as well. The other thing was that the officers themselves, many of them, particularly in the South, were increasingly coming to view that they would be better off if they were self-employed and that they could pay less tax, or defer tax or sometimes there was no tax. The Government, the successive government, then began to respond by introducing a scheme to try to ensure that these people did pay some tax. There were all sorts of different schemes, according to CIS, but basically, they were all to try to establish what a man's tax system was and I have to say that most of them didn't work very well and I think they won't because the industry is a transient industry, people move about and stuff'.

This excerpt demonstrates two critical points. First, the legacy of government action transcends beyond political ideology. The growth in self-employment has largely been blamed on the market philosophy and individualistic enterprise culture introduced by the then-Conservative government under the wing of Margaret Thatcher. Yet, this reflection suggests that the legacy went further back to a Labour government, which sought ironically to tighten the tax system up so that those who were not paying taxes in the informal construction sector became more visible for tax purposes. What started as a Labour policy, in spite of differences in political ideology, was then modified and perpetuated by the Conservatives who took over in power. The second point worthy of mention is that there are often unintended consequences of any intervention, government legislation included. Sir Michael Latham reflected on the ramifications of the selective employment tax: 'But, since 1966, there has been a very substantial movement towards self-employment and many of these self-employed people, of course, are not actually self-employed at all. They are working in gangs for a labour master, a gang master, or a labour agency. And there are now hundreds and thousands of them and in the South of England, that's all with agents. When you go North, when you go to Scotland, for example, there is less of it. And in 1966, there was virtually none of it. But that's no longer the case. There are still plenty of people up there who are directly employed, in Scotland and in the North East and as a result, there is much more training done up there than there is in the South of England. But, self-employment has spread there, you know, they're doing the gangs and also, I have to say, has imported labour'. So government legislation does have consequences, intended or otherwise, in relation to shaping the nature of the construction workforce and this has a significant bearing on the sustainability of employment practices and workforce development.

So what can governments do to secure the longevity of the sector and deliver on sustainable development?

Nick Raynsford asserted that government officials need to really get immersed in the intricacies of the sector for which they are responsible if changes are to be brought about that could have a real impact. He felt that when he was a Minister of Construction in the UK, he was provided with adequate levels of resources in time and staffing support that enabled him to represent the industry's interests effectively. He surmised, 'I was very lucky because I had four years in opposition and then four years in government working on the subject. And in eight years, you not only develop lots of knowledge and understanding, but you can also build a network of contacts. And I certainly did that. And having a four-year period, you can see through all the innovations you've introduced. So, the Egan report we commissioned in 1997, it was published in 1998; the Movement for Innovation was set up in 1998 that led on to a whole series of demonstration projects, which we oversaw. There was a lot of work done to try and engage different sectors of the industry: the clients, the main contractors, the professionals, the subcontractors, the specialists. And then we introduced new structures that were designed to carry forward the whole reform agenda. And there was continuity there. And I think my successors have suffered from relatively short terms in office'.

Stef Stefanou also considered continuity in government to be important as this would help maintain a wider, long-term view on what value really meant in terms of the provision of sustainable development. Stef was both frustrated and sympathetic about the role of the government, as he commented, '[the role of the government] is supposed to be changing. But somehow, no matter what they say, they always go for the lowest price, or mainly for the lowest price [. . .] but the way they do it and feed it to departments and the departments don't change, because they have so many departments, and you have to understand it from the civil servants' point of view. Why should they risk? Why should they risk their house, their savings and everything in order to take the best value contractor? And then after five years, somebody accuses them, you know, the general office of audit says they've done something wrong and they are being sued, like some councillors etc.'.

So, whereas Nick Raynsford reminisces the time when there was relatively more resources to staff a department that looked after the industry's interest, Stef Stefanou recognises the changing departmental structures in government, which, in turn, bear implications on the way the idiosyncratic nature of the industry is being understood and represented. Such constant reorganisation of government departments diminishes the ability to hold a long-term view and subsequently implies a greater need for coordinated thinking and action across departments in government, as highlighted by Stef Stefanou. And, of course, in the last decade, where a contradiction has developed in terms of devolution of political power to the regions and localities on the one hand and relinquishing of authority to European governance machine on the other, the issue of joined-up thinking in governance has become even more pressing. These dynamics will be further explored in Chapter 4. But what are the consequences of such shifts in the way the way political leaders frame their focus?

It's a numbers game: how governments devalue what construction does

It all becomes reduced into a numbers game! Government policy is often framed in numerical terms. Specifically on the relationship between government and the construction industry, the target-driven culture means that the focus becomes centred narrowly on the financial costs of construction, without ascertaining the true value of the built environment. Bob White, when describing the huge public sector programme *Building Schools for the Future*, was somewhat sceptical of the government's efforts to improve the educational infrastructure through greater involvement of the private sector. According to him, the aspirations of raising educational attainment outlined in government rhetoric are not necessarily met by the mechanisms of delivery in place. He argued that such an agenda should be much wider than the provision of modern buildings, thereby returning to the points made above on the interaction between people and buildings and the community.

He insisted that the role the government plays in demonstrating political leadership is crucial if aspirations such as raising educational attainment are to be met. Bob stressed that there needs to be greater involvement of the public sector in overseeing the financing and delivery of education improvement programmes that move beyond the building of schools. Furthermore, he suggested that simply devolving the delivery process to the private sector is insufficient as decisions will be made purely on economic terms, and, sadly, he noted that 'the Government often undervalues the cost of building' anyway, which potentially threatens the realisation of policy aspirations.

We will explore the idea and consequences of the apparent relinquishing of public responsibility in infrastructure provision on to the private sector in the next chapter. But the issue of undervaluing the true costs of buildings also touched a nerve for Chris Luebkeman: 'Some of the things that cannot be measured, some of the values. . .how do you value a St Paul's Cathedral? How do you whole life cost it? Easy. What about the whole life value? The value to a city. The value to a culture. The value to a place in the minds of people. And so, whole life valuing needs to be something that we can articulate as an industry'. The emphasis on quantitative measures in policy-making only promotes a myopic approach to creating outputs, some of which might not matter. For Jon Rouse, policy-makers ought to be focusing on the more qualitative, much harder-to-measure, outcomes that any policy creates. Jon explained, 'If you take the post-war period, we started really in the 1950s, it's all about people. If you look at Abercrombie and his maps and also things like the Barlow report on the issue of new towns, it was all about influence and drawing lines on maps; you need radial roads around London, you need three or four routes in terms of rail around London etc. It was very process-orientated, very mechanistic. I think in the 1980s, particularly following the social riots in Brixton and so on, we moved much more to an output culture. So we got to reduce unemployment, we are going to measure how much we are reducing unemployment by and we've got too many houses that are in disrepair and we are going to reduce that by this number. I think, in the late 1990s, with this Government coming in, we genuinely started to see some of these outcomes. Now, an output simply is a numerical measure, so somebody who's unemployed that's got a job. Someone interested in an outcome would actually ask: "what job?". So if you look at the rate the corporation has grown, we're still stuck on outputs. The key thing they used to drive the corporation was: how many units were built last year? I can tell you. I mean, we built 24,200, but, you know, do you want to know anything else about them?'.

So understanding and then enacting on value, both measurable and intangible, has got to be critical when political leaders consider interventions for sustainable development. Yet, as we shall see in the latter half of this chapter, knowledge about what sustainable development truly is remains debatable. And so, as Bob White argued, there is a need for deeper governmental involvement in that debate if the wider agenda of sustainable development beyond narrow, economic and numerical targets is to be achieved. After all, the construction industry cannot be left to its own devices to meet sustainable development in the wider context. As Bob noted, 'One of the problems with our industry: we build, then it lasts too long by the way. Everyone says it's sustainable. So what does sustainable mean? I mean, the unfortunate thing about this is, even when you're talking about the good things about industry and progress, you're only talking about costs. Much of the industry is stagnant and sterile and lacking in any sense of service to the public at all'. He added that the construction industry may be instrumental in providing the physical aspects of infrastructure and the built environment, but the sector would benefit from a greater level of government steering and prescription if it were to meet some of the other social and human needs of sustainable development.

Communication of government intentions: clarity or conspiracy?

Yet, governments are not very good in making their intentions transparent. Up to now, the environment and the increasingly important low-carbon agenda have escaped much of our reporting here. It is clear that apart from the fears of energy consumption, the low-carbon agenda certainly featured less prominently at the time of undertaking the interviews. Without a doubt, this has now grown to become a globally more prominent issue, especially with the buy-in from the Obama administration into the climate change protocol in the USA and their endorsement of the generation of green jobs. However, Stef Stefanou questioned whether we fully understand the science behind the low-carbon agenda and climate change, as he suggested that the debates surrounding sustainable development can sometimes be hyped up in the media and policy circles. Perhaps Stef has good reason to remain cynical, as he outlined a few examples of where evidence was less forthcoming in the policy assertions in a number of areas. 'Some of [the policy-makers] have never been on a site and then they create a problem from the industry because they develop all these initiatives that do not make sense. They make sense in theory, but not in practice. You remember everybody from 1996 onwards was telling us when the clocks reach 31 December 1999, the whole world will collapse and that businesses will collapse because of the Millennium Bug. They wrote report after report after report by these consultants who convince three-quarters of the world that this was going to happen. I think the British government spent about £18 to £20m to run seminars, conferences, ministers making speeches, and you can find all these speeches on leaflets, pamphlets. It was going to be the end of the Earth and at the end, it was a red herring. It was a result of consultants, of experts etc. It's like the world will burn tomorrow according to these people. And it's amazing that nobody even challenged these people after the event, to say, "Hey guys, how could you have been such experts when nothing of that happened?"' Stef was making the point that good intentions by government must be matched with transparency and accountability. Lack of clarity can sometimes lead to confusion and a feeling of despair that government initiatives are simply conspiratorial.

To illustrate with another example, Stef bemoaned the low-carbon agenda, suggesting, 'Now, we have all these experts in CO_2. The whole world will burn and become kebabs according to them, but I don't think so. Ok, the temperature rises, but then if you go backwards in time, you can find chunks of periods of time when the temperature has been rising and it has been dropping and rising. If you go millions of years back, I don't know, I am not a scientist; you know, there are a lot of academics who say they can prove it, and also [. . .] there are a lot of papers as well, especially by the Swedish professor, Bjorn Lomborg, who proves the opposite'. Stef is probably a typical practitioner who feels that he is ill-informed by policy-makers (and experts) in terms of their real intentions, yet he seems to be sufficiently reasonable to consider both sides of the academic debate i.e. that of the climate change proponents and that of the sceptics and deniers. However, scepticism will only lead to the consideration of any intervention as just another revenue-generating mechanism, as Stef added, 'I mean what's the point in saying we are trying to save the Earth in 200 years' time, when you are letting people die now? That money should be spent now. I mean the child or family in Africa who is dying, these people, what do they care about CO_2? Now, why do the government encourage this, consciously or unconsciously? Well, the funding system has to work somehow, so almost the monster feeds itself. The other thing is if you are the government sitting around, scratching your head, thinking, "How are we going to get more money to run the country? We cannot put taxes up because of the election, we have already dealt with that tax increase there etc." So, some bright guy says, "CO_2, we are going to do something, let's tax CO_2, we are going to get more money, and we'll be perceived to be doing something"'. So, it seems that before an industry response can be coordinated to confront the sustainable development agenda, there is a communicative requirement for greater clarity in government policy and rhetoric so as to secure buy-in from practitioners like Stef.

Governments can encourage actions at the grassroots level

Yet it is simply unfair to say that the vast majority of industry practitioners and individuals would simply rest on their laurels when it comes to tackling such grand challenges. Some of our interviewees maintain that the industry can afford to, and often, do more as part of their social responsibility. George Ferguson said that it is time for the industry to mature in its approach to dealing with wasteful practices: 'And I think we need to grow up to understanding that we shouldn't just use everything because it's there. Like, being able to build a 500 metre high building, you know, or whatever. It's so not necessary and not sensible just because we can do it'. Chris Luebkeman echoed this sentiment as he suggested that a deeper understanding of what constitutes waste needs to be harnessed in the industry, that there is a need 'to understand that waste is a misallocated resource. So we, as an industry, can really sharpen up the waste, in terms of resource, that includes building material creation, building sites, the running of buildings. There is a desperate cry around the world for a way our world is going to be when we run out of oil, when the seas start rising, when, you know, when the world is a warmer place. There is a desperate cry for some visions out there that we can move to with'. So these extracts suggest that our leading figures are indeed thinking of the wider, environmental impacts that construction activity brings, and that governments can do more to reward good practices or penalise wasteful processes.

For others, there is also a lot that individuals can do for themselves to create the sustainable communities in which they live. Tom Bloxham suggests that one must never forget human resilience in sustaining their survival prospects. And in the contemporary context of the first global economic crisis in the twenty-first century, economic lessons learnt from the past regarding decisions and the entrepreneurial spirit remain valid if we are to understand how we can sustain communities. Tracing the growth of Manchester, and especially focusing on the adaptation of buildings, Tom Bloxham noted that buildings like communities do go through periods of renewal: 'it was more a business thing, about supply and demand. You know, there was this demand from the entrepreneurs and there was supply and because the property market really just crashed around when the Thatcher government was encouraging enterprise. So to buy buildings, you could do it cheaply as long as you had the entrepreneurial drive. If you were setting up business – and a lot of people said to me, "Well, that's £40 a week of my allowance scheme" and there were a lot of people, for the first time, said, "'I don't want to get a career. I want to go into business". Anyway, alongside that was a decimation of the manufacturing industry. So, there were no long-term jobs. But also, all the space that the manufacturing industry had traditionally used, then moved out, leaving the buildings empty and they were a liability rather than an asset. So it was cheap to buy the buildings and adapt them'. For Tom, enterprising individuals who seize economic opportunities from adverse situations do much more than survive; they potentially revitalise the landscape of society. But, central to this shift is a Conservative government who was encouraging such entrepreneurial drive. Therefore, governments have demonstrated the ability to influence human behaviour at the grassroots. Shifting to a modern-day equivalent, the acceptance of the need to create green jobs can be seen as an example of how governments across the world are seizing an opportunity to try to recover from the global economic recession. What remains to be seen are the consequences, both intended and unintended, of such policy shifts.

Nonetheless, there is much scope in a recession to think about opening new market opportunities. Tom, for instance, reminded us that infrastructure development is not simply about newly built facilities. The adaptation of buildings also translates to 'recycling' of physical spaces that could potentially save a lot of resources put into building new facilities. His own story illustrated how his eye for property development, together with the desperation of new enterprising traders to find a space for doing business, have contributed to him identifying a business opportunity in the adaptation of physical spaces: 'We're now talking late 1980s. [The entrepreneurs] had nowhere to actually trade from. Take Camden Market, you went and put up a stall there, and then it becomes a market. That sort of thing, that's how it actually started. At that time, the office type places, as it happened, was the only way they'd be allowed in. But, because they would never want to sign property leases, so the only way to have them in, is buildings that people had demolished and didn't know what to do with them. They'd be very poorly maintained, they'd be on short-term licences and they'd be in there for a year or two, pending a re-development or something. They'd get moved round from building to building in Manchester. So, that, alongside my own need for space, I had a company going, looking at retail space, and I said, "'Well, actually, let's make use of some of these open spaces". So, we started doing it with Afflecks Arcade and went after the space needed for our poster business, we had bit of space left unused, so we then sub-let it to other people and it snowballed from there'. Thus, this demonstrates the power that people have to challenge the status quo and produce something positive.

Wayne Hemingway also believes that economic decisions determine the extent to which people adopt green practices: 'We lived in Morecambe and then we lived in Blackburn and you know, life was spent walking to visit people, walking to the shop. We never had a car. [. . .] Until I was seven, until my mum got married, we didn't have a car in the house. And even after that, I don't think I ever got a lift to school or anything. I was always walking to school, walking to the shops and walking to play football and everything and so, you are bound to have more of a sense of community when you're doing that because everything is based around what you can get to within a reasonably short distance'. According to him, it was not because of his environmental consciousness at the time, but out of economic necessity that he had to rely on his own two feet to move around. Of course, Wayne Hemingway is now instrumental in designing communities of homes that try to re-create this sense of community where everybody knows everybody else and where services are within close walking proximity, which removes the need for a personal car parked up by the porch. It is unclear whether Wayne's experiment of 'forcing' home-owners to ditch the idea of having a car parked outside their homes has worked. Nonetheless, it is Wayne's belief that the economic imperative, although potentially detrimental for its narrow numerical focus explained in the preceding subsection, can at times be beneficial to mobilise decision-makers in industry and people in communities to behave less wastefully. Yet, framing policies to facilitate such actions require a deeper understanding of the complexities of human behaviour, which demands adequate resourcing by political leaders if a longer term view beyond the crafting of consultancy reports is to be held.

To summarise, therefore, we have identified the importance of infrastructure development in developing sustainable communities, and the role governments can play in this regard. Our interviewees saw the role of governments as major clients and regulators of the industry to be critical, as behavioural change can be engendered through their procurement policies, a wide range of legislation, greater clarity and transparency of government intentions, and a better understanding of how human behaviour can be shaped by policy-making. There is, however, no prescription here for what must be done, because, again, it is clear that economic, social and environmental concerns are enmeshed together in a rather complex way. That said, the dominance of the economic perspective is both a blessing and a curse, as such a perspective often leads to the mere numerical framing of public policy and the devolution of responsibility to the private sector, but, if marshalled well, can be a very potent force in engendering behavioural change among people in corporations and communities.

We will return in greater depth to the dynamics of governance in Chapter 4, but the key messages on the role of government in developing infrastructure for sustainable development are threefold.

- First, the political focus hitherto has centred narrowly on financial costs, thereby underestimating the true value of the outputs and outcomes produced by the construction industry;
- Second, some of our leading figures have expressed their frustrations and cynicism about government intentions on the sustainable development agenda and that this is not helpful in moving the agenda forward. Nobody really fully understands what this agenda is anyway; reinforcing the incomplete knowledge that society has to muddle through;
- Third, the industry and individuals can still play their part in reducing wastage and encouraging green practices, and governments can do more to understand the complexity of human agency to design policies that incentivise/penalise behaviour accordingly.

Clearly, there is more that needs to be done, both in terms of a deeper understanding of what sustainable development really means and also the institutional coordination and response to this agenda. This will be the focus of the next chapter. In the next section, however, our leading figures inform us what the industry response could be in the future.

Industry response to the sustainable development agenda

As mentioned earlier, notwithstanding the importance of government policy and regulation in shaping the enactment of the sustainable development agenda, there is a role that practitioners in the industry can play in terms of making incremental change happen. In this section, our leading figures talk about the need for industry to reflect on the wasteful practices and urge practitioners to do all they can to stamp out waste across the supply chains. In so doing, our interviewees suggest that savings made can be ploughed back into investing in research. However, there remains the problem of short termism in the sector, which can impede progress made on the sustainable development agenda. This is unlikely to disappear as it requires shifting attitudes over generations through education.

Chris Blythe considered the need for the industry to reflect on its wasteful practices: 'Thirty per cent of the industry's effort is waste, if you just tackle that, you know, you don't have to change very much in your business'. Guy Hazelhurst also suggested that what is required is for the industry to reflect on current practices and to adopt greater integration across the construction supply chain and their users. He explained, 'Firms get work on local jobs. If local firms can't get local people jobs, then you're not going to build the local skills capacity. One of the key issues that you hear much about is that it's the same local labour initiatives in Liverpool when they were creating the city of culture in 2008 or East Manchester for the Commonwealth Games in 2002 or whatever it may be. Everybody voted on the availability of skills to work on regional projects, but they missed the fact that local firms need to provide the work to do so. Take cladding, you know, glass cladding. I did a project in Bristol which showed that the nearest firm where you could get the work done by was not from Bristol but Southampton. Now that company is not going to employ, generally, Bristol people. . .They'll bring them from Southampton. They'll come to the project site in the morning and they'll leave in the evening. And yet if you look at the Bristol phone book, there are Bristol cladding companies. The thing is it's not got the right type who can hang that kind of system on that kind of building'. Resolving this by thinking about the supply chains when producing specifications at design could, therefore, contribute to a greener approach to construction.

Indeed, Chris Blythe suggested that it was instrumental that practitioners remain self-critical and reflective about the efficiency of their operations, as he argued that the savings made could then be returned back into finding practical solutions that make the sustainable development agenda less elusive. He suggested, 'If you split up, 10% for the firm, 10% for research and 10% for investment, you know, you have a very healthy business. The firms that have the major savings can succeed further, if you see new ways of producing, new ways of manufacturing and new ways of construction. It's all there to be got. The issue is, do you go for it or is everyone too comfortable with what we are doing? But, there are drivers that will change things'. One such driver, of course, is the current economic recession, which has paved the way for the creation of green jobs, at least on a rhetorical level.

According to Chris Luebkeman, more work needs to be done in terms of basic research to come up with ways that can help produce better environmental solutions for the future while ensuring that the wheels of the economy remain greased. Chris suggested, 'The world will be a warmer place and the seas will rise. We are going to run out of oil, even if we stop pumping CO_2 into the atmosphere, the world is going to be warmer and no one is going to stop pumping CO_2 in the atmosphere because no one is going to let their economy fail. It's sad, but it's simple. You know, I don't want to be out of a job, right? Nobody does'. However, he warned that 'for the next ten years, on a global scale [. . .] those ten years are going to have a fundamental impact on the following century. Fundamental! And we're already in to that now. I mean, this decade, these next ten years are going to impact the following ten years, so much more than the Seventies impacted on us. I mean it took us 135 years, you know, to use that first trillion gallons of oil. It's going to take us 20 or 30 to use the next trillion [. . .] It's very scary'. Chris suggested that 'we've got to get our heads round the fact that, you know, we've done a few things to the world and now we are going to have to deal with it. So, we need to start really getting our heads round that and try to imagine what we're going to be doing to retrofit buildings and spaces and places that were designed based on last century's climatic rules, not next century's. That's real research'. Ultimately, for Ken Yeang, knowledge generated from such research needs to be embedded in the training of future professionals through what he termed as 'the green curriculum', although he conceded that there are no quick fixes here: 'It will take 20 years for that to happen I think. We need to change the curriculum, we need a whole new breed of educators that will embrace the green curriculum and I see this happening more and more as schools of architecture become more conscious of the need to build greener, but they do not know how to do it yet'.

However, the industry often lacks the tenacity to think over the long term. Chris Luebkeman suggested that one of the critical challenges that confronts the sector is the need for a 'very deep understanding of whole life valuing, not whole life costing, but whole life value'. However, there are often tensions between taking a long-term view about issues and the ability to subsist in the meantime. Unfortunately, Chris Luebkeman argued that whole life thinking is 'something that [the industry and society] don't have yet'.

Nonetheless, for Stef Stefanou, industry resilience and the ability to adapt to the changing environment is what industry can pragmatically achieve, even if the response constituted short-term fixes rather than thinking about the longer term. He recounted another recent example of the fuss surrounding the notion of skills shortages, and suggested that the industry managed to cope in a somewhat adequate manner: 'I think everybody for years have been telling us we are coming to a standstill because we won't have any trades, enough trades to do any work. However, life is not like that. It's like people in the past, in the early twentieth century have been telling us that industry will come to a halt because we will not have any coal left. Yeah, that never happened, things change. So, the shortage of trades. And then suddenly Europe opens up with 10 more nations, and now another three or four. And now, the trades have been supplemented by these countries'. For Stef, the utilisation of migrant workers has been good for the sustenance of business, although he did concede that the industry must also get its act together to ensure that capacity continues to be built: 'I think we have been provided with a window of 3 to 5 years to start training these trades to bridge the gap of this shortage. And the reason I am saying

3 to 5 years, because I am talking to these people and everybody I have spoke to, they want to go back to their country within 3 to 5 years, all of them! Obviously, some of them will fall in love with an English girl and stay etc. but the majority of them want to go back to their country'. Therefore, it is imperative that the industry considers sustaining its capacity in the future. It is here that we turn to the role of education and research.

Role of education and research

As was mentioned above, the sustainable development agenda has come of age since its theoretical conception in the 1970s. Yet, it has taken three decades for this agenda to enter mainstream discourse. Indeed, it takes generations before attitudes and behaviours can be altered so that the population at large can act more sustainably. It is, therefore, not surprising to find that our leading figures accept that there is a role the education system can play to deliver such changes. In this section, a number of critical issues raised by our interviewees are discussed. A point is raised about the inherent contradictions faced when addressing the sustainable development agenda. There are tensions between the desire to maintain the long-term view for the often speculative benefits for future generations and the immediate, economic concerns driven by short termism. Therefore, our interviewees place credence on the need to integrate the younger generation through the education system. On the one hand, this ensures the sustainability of skills for the future of the industry, and more critically, this will allow fresher perspectives to be developed. Yet, there remains one crucial problem that needs to be resolved, and that is the need to encourage diversity, whether this is in terms of the make-up of those who enter the industry or diversity of interdisciplinary knowledge deployed in the sector. In any case, our interviewees see much benefit in forging closer links with the education system and suggestions were made for a partnership model that involves the social partners of the state, employer and employee representation. Ensuring closer ties between industry and academia appeared logical given that the education system, through its teaching and research activities, can contribute much to the continuous improvement agenda, especially where knowledge about professional management and technological development are concerned.

Inherent contradictions in sustainable development

Any discussion about sustainable development is always confronted by the dilemma of bridging the intergenerational gap. How do we engender future thinking in such a way that challenges the status quo at present, while coping with a sense of pragmatism in the short term? For Alan Ritchie, he was critical about both the quantity and quality of skills to safeguard the future of the industry. He noted the importance of building future capacity as he stressed that 'apprenticeship and training I value very highly for the industry'. He added, however, that there is a crisis in terms of capacity building, in part due to the ageing workforce particularly in the developed world, and in part due to the lack of institutional support in determining the nature of skills in the sector: 'I can't understand their logic, a billion pound industry and yet, we can't determine who's all coming in. There's no monitor of that skilled labour to develop the industry five years or ten years down the road.

That's why you have an ageing workforce. I think the average age of the workforce in the industry is over the 50s now and being the very nature of the industry what we are going to find is that we are going to be retiring very quickly and unless we start developing these skills, then the industry is going to pay for it'.

However, apart from the quantitative problem of skills shortages brought about by natural attrition, Alan Ritchie was also concerned about the quality of skills being developed. He argued, 'As a trade union, we are not interested in just short termism. What we are saying is, as the trade union, we are interested in the long term of the industry because they are going to be our future members. And I remember having a discussion in Scotland with an employer and what he done is, he fitted double-glazing windows. And he had an apprentice joiner and we were arguing about the wealth of skills that joiners should have. He should learn dove-tail joints, mortice and tenon. You know, developing his skills on how to hang a door properly. And this employer said to me, "Alan, I'm not interested. I'm fitting double-glazing windows. I'm not interested in him learning anything else, as long as he can fit double-glazing windows". And that's the short-termism that you'll find with some of the employers'.

Such emphasis on immediacy does not align well with the ideology of basic research, which Chris Luebkeman argued is necessary for advancing the industry. He presented the tensions between the purpose of academic research and the commercial imperatives of practice, 'In the academic world, one has and one should have time to think, to think deeply, to investigate and to share the challenge with a generation who hopefully is dissatisfied as every generation should be. Hopefully, every twenty-year-old in the world is dissatisfied with where they are. You know, challenging where things are. And, as I say, to me, academia has a very important role to help with this challenging of the status quo constantly. Right? There's this deep thinking and researching into areas which are vague. They should bring that back out. So, there's always a bit of, to me there's this sharing of, of pure focus time. And academics should have focus time in various subjects of interest that they can then be a researcher, begin the ripple effect with their researchers, to challenge and be challenged and investigate and like a big ball of clay, rip it apart, put it back together, rip it apart, put it back together. At the same time there's this steady stream of twenty-something year-olds who are challenging the new concepts. Because the context is not static, it's always changing. And a new context is evoked through new students, because new students, their age stays the same, but their context is constantly varying. And that is a crucial role [of research] because in a corporate world, we are rarely confronted with questions that you hadn't thought about'.

However, often bringing in new, fresher generation of ideas can be easier said than done. The industry is well known for blocking the recruitment of non-traditional sources of new entrants, for example. Sandi Rhys Jones bemoaned the failure of the industry to attract new entrants, and especially women, into the world of construction work. However, this is not entirely the industry's fault, as she explained, 'One of the things that continue to distress and depress me is the poor or the unimaginative careers advice that is available in schools to young people. It doesn't seem to have progressed very much and certainly, when I was at school, construction engineering, built environment, any of those areas, simply weren't on the radar screen. I don't think we even looked at architecture. At my grammar school, our career paths were very clearly mapped out according to our perceived particular skill and competence. So, I was clearly labelled for all my school life as a linguist, classicist, writer, and so on and there was a very clear separation between arts and sciences'. Although the industry might be improving in terms of encouraging diversity, there is still much room for improvement in this respect.

George Ferguson also echoed the importance of diversity and welcomed the notion of retaining ideas from the younger generation while drawing on the benefits of a variety of experiences from people of different backgrounds, 'I have always had an emphasis on youth and thinking that, you know, if you can get it right in the early years, then you get a better educated, more socially aware, more artistic view, you know, more visually aware population. So, I have always felt disappointed really that we are such a visually illiterate nation and that we haven't moved as fast as we should have done to being a visually literate nation, like I think some of the other European nations are. You know, education tends to give us more literal literacy than visual literacy, I think. Anyway, being a councillor was a fantastic experience because I did, for many reasons, get very close to people whose lives I would not have understood otherwise – people from very different backgrounds'.

There is, indeed, more that can be done to bring the different academic disciplines and industry practice closer together. Whereas Jon Rouse finds the time to engage with the education sector and academic research, he considered the issue of relevance, as he remarked, 'I love knowledge, I love spending time with academics and I go out of my way to do stuff. So, for example, I sit on the Joseph Rowntree Foundation's research committee on planning and housing and I also spent a number of days last year to be involved in York social housing group. So, I love taking time out to pursue knowledge, and I love taking time out and spending with academics and I could see myself as an academic quite comfortably. My only criticism of academics [...is that] there is too great a level of detachment between the practical needs of government and industry. Too many of the research questions that churn out answers that have no practical impact in terms of the questions that are asked in the first place. I am amazed how rarely an academic, a group of academics, an academic body comes to me in my career and said, "Jon, what would be useful to you?". You know, "If I think about a research project over the next twelve months. If I'm thinking of putting something together and thinking of what it might mean, what would actually be useful to you at the corporation?". You know. "What would actually make a difference?" I could count on one hand the number of times that academics have actually come to me proactively'.

Getting education closer with industry: towards a partnership model

The need for joined-up thinking, however, extends beyond connecting the education sector and industry practice. On the skills agenda, Alan Ritchie considered the need for supporting a social partnership model, where government bodies and the industry represented by employers and employees engage closer together. He highlighted the case of Scotland, and suggested that forging closer dialogue between the social partners can only be a good thing in terms of building up future capacity, as he surmised, 'in Scotland, there's only Scottish Vocational Qualifications Level 3 they recognise as a craftsman, whereas in England and Wales they are saying, you know, "We'll recognise Level 2". And it seems to be trying to deskill the industry and that is a worry because, I mean, where do you go? [. . .] Now, in Scotland, you could never complete your apprenticeship just by going to college. You must have an employer and you must have experience, on-site experience before you get it. But, the other most important thing is: the joint awarding body in Scotland is the Scottish Qualifications Authority and the Scottish Building Apprenticeship Training Council. That is a joint awarding body, whereas down here, it's with the Construction Industry Training Board (CITB) and the CITB is employer-led'.

Developing joined-up thinking and working and getting engagement from industry actors is certainly an important point that we will return to in Chapter 4. However, our interviewees recognised that this is what the industry is doing poorly. Bob White suggested that a major segment of the industry, the small and medium-sized enterprises (SMEs), sometimes do not engage with cutting edge thinking simply because it is often easier to remain in the comfort zone of the status quo: 'How are we going to contact SMEs and get them aligned with this? Who cares? If they don't want to join the party, then, you know. . .I was an SME once and I chose to do that. Some SMEs just do it for ease. And, it's a lifestyle, you know. You get a local architect who I don't know, knocking in somewhere and he's got ten people working for him and that's for his own benefit. It's not for anything else. It's not to further the profession or improve the industry or anything like that. It's a lifestyle thing and if that's what you want to do, then that's fine. But, don't then let them get the RIBA all upset because they're not being approached by all this innovation stuff'.

Yet, it is often assumed that cutting-edge thinking has to be embraced across the board, regardless of the context in which a variety of firms operate. However, much innovation is often conceived with larger firms in mind and fails to consider the heterogeneous landscape of corporations, especially where SMEs are concerned. Instigating change across the board through a 'one size fits all' approach might therefore be inappropriate, and, at times, simply promote the idea 'big is best' at the detrimental expense of smaller firms. Bob White, drawing an example from the retail sector, noted 'I remember when the big stores emerged for the first time. You know, your Sainsbury's, or whoever, there was all this complaints about what's happening to the high street and what's happening to the small corner shop? And how are we going to make them sustainable? Well, [the small corner shops] no longer exist now. Everybody is used to going out of town to shop and everybody enjoys it as a leisure thing and who cares? You know, that's what people do'. He added that 'What people want is to live in a good environment with, you know, good education, relatively good food, you know, lifestyle and security and all these sort of things. They don't really mind what the company size is that delivers those components'.

Thus, both George Ferguson and Bob White consider the engagement with SMEs to be especially challenging for driving change in the industry, yet their responses are somewhat distinct. For George, striking partnerships with SMEs can be an aspiration to achieve, but one needs to better understand what motivates the proprietors of many of the smaller sized firms that exist in a rather hand-to-mouth fashion. From another viewpoint, Bob prefers to place the focus on what the end consumer wants, and despite the excerpt above, SMEs do have markets that they serve. It is, therefore, vitally important, for Bob, that the needs of these consumers are appreciated and met. So, in engendering change towards a more sustainable future, education undoubtedly plays a critical role. In this subsection, it is explained, however, that a 'one size fits all' approach would be inept to cope with the heterogeneous landscape of stakeholders involved. As the examples demonstrate above, this requires forging closer partnerships, whether this is between the social partners of the state, employers and trade unions in shaping the education system, or a better comprehension of the multitude of subsectors and variety of firms that deliver construction goods and services.

Continuous improvement in the sector and the role of research

Continuous change for the better certainly featured prominently as a theme for our interviewees. Bob White, for example, suggested that the industry is actually very good at doing things [. . .] to help produce stuff that others get profit out off'. Yet, Bob argued that the industry can sometimes be made to feel isolated with little support in terms of promoting this good practice and tackling grand challenges of the future. In particular, Bob White was concerned that the industry cannot simply function on its own, and saw the need for partnerships with government. Specifically, he identifies a number of big issues that needed to be addressed, which require support from these partners, 'There are three or four big issues sat here waiting still to be cured. Not least of them is diversity, you know, and equality and why haven't we got enough women in the industry? Because if we got women in the industry, it would cure the industry ills overnight, because we get capacity and because we don't seem like sensitive normal people in the world and you know, the macho stuff. Secondly, you know, why haven't we got proper skills? You know, why haven't we got a decent education system that recognises exactly the sort of people we need and trains them for those things, instead of training them in all sorts of stupid stuff? Why haven't we got proper representation in the industry for, you know, with people like government? Why don't the ministers do their job properly? Why don't the agencies that have been set up to do their job do it?'. For Bob, there is a definite role here for greater government support in these areas.

On the contrary, Tom Bloxham argued that although the industry is making progress over time, there are still a lot of ill practices that need to be stamped out. In a similar vein, however, Tom suggests that there is a role for educating the industry in professional management to improve the current state of affairs: 'This is an absolutely crap industry. It's getting better slowly. I think two things have affected it. One is the knowledge of the main contractor. You know, how do we project manage it? So, I mean the whole basis of it being adversarial, yeah? And whatever people tell you, the first thing is the construction industry takes a view on how difficult a scheme may be, and they'll bid on it and even minus profit, knowing they can get a claim in at the end of it and therefore, the best paid people there are not trying to fulfil a role, but trying to get claims back at the end of it. Terrible, terrible state of affairs, but it is improving. It is changing. You know, project management, management of change, the adversarial nature of the industry. The whole basis of it, the whole purpose of the construction industry is to build, just enough, not getting better. You know, which is really sad and actually, I think, in most cases, people want to do the same thing all the time. There's don't always invest in technology, whereas every other industry, you pay for technology, you pay for the production and you pay for the materials. There is still a traditional quality in construction'.

Indeed, Sandi Rhys Jones echoed this sentiment that the industry can at times be backward, as she suggested that the construction industry 'has to find a way of making itself an aspiration in society. We need more people who want to do it, rather than find themselves doing it. It has everything in place. It needs to find a way of improving the way it manages people. How do you find a framework for people joining the industry to see some form of progression?'. Of course, there is much scope for the education system to be an integral part of delivering the improvement agenda outlined in this subsection. On the one hand, applicants from a wide range of backgrounds can be further encouraged to undertake construction courses so that the next generation of people working in the industry can reap the benefits of diversity. Furthermore, there is the opportunity of applying research knowledge on modern aspects of technology and management already found in universities to aid progress on the continuous improvement agenda.

For Stef Stefanou, engagement by the industry actors has a more pragmatic value. On his thoughts regarding industry's involvement in basic research, Stef remarked, 'I have no problems with being as close with Universities. Of course, when the company was smaller, it was not possible because we had less people and I had to run here and everywhere. However, about 10 to 15 years ago, I woke up one morning and thought, "We are criticising all these documents that are coming out of universities, and sometimes these documents act as the basis for governmental and departmental policies". And I thought to myself, "We do criticise, we talk to each other and say how useless they are because they are theory, and yet, whenever university graduates or postgraduates or professors send us letters asking us for feedback or asking us to participate, we say we haven't got time and we throw them in the bin". And then they prepare something theoretical and then we complain. And I say, "This is wrong! If we are prepared to criticise and expect something more solid, we have to contribute". We have to engage at that time of the thinking and explain our point of view, rightly or wrongly, at least it will be ours, and influence them on what they are writing; not try to do it after they wrote it and just criticise them. And that's how I try to get involved and I encourage my guys to get involved'. So what Stef has pointed out here is that there is much benefit for the industry to co-produce research outputs with the academics, returning to the point made about the need for forging deeper partnerships between industry and academia.

Summing up the thoughts of our leading figures

To sum up thus far, sustaining the longevity of the industry featured as a prominent theme for our interviewees. There was acknowledgement of the contribution the construction industry makes in terms of underpinning the productive capacity of other sectors, and for supporting the livelihoods of people in communities. All our interviewees recognised the importance of people when framing their thoughts about the future. After all, physical structures of the built environment would be meaningless if not for the people that design, construct and use them. Yet, the social interactions between people and buildings are interwoven in a complex manner through what our interviewees describe as communities. Although many refrained from the use of the phrase 'sustainable communities', and indeed there is no prescription here of what is the best approach to sustain communities, our influential figures considered the social dimension of how communities helped shape the built environment and vice versa as a critical factor that needed more comprehension when discussing the future of construction.

Yet, when explaining their views on sustainable development, there were a number of contradictions that developed in the conversations. For one, there were tensions that emerged when talking about the future in terms of time, as interviewees struggled between the pragmatic need to meet short-term objectives of satisfying current demand and the prevalence of the profit-making motive on the one hand, and the desire to keep a long-term view for the benefit of future generations on the other. This was certainly a source of contention and uncertainty when interviewees talked about concerns about the environment and the climate change agenda. The views expressed here were also divided; whereas some interviewees recognised the problem, especially in terms of the future supply of energy, others remained sceptical about the way interventions are being devised to tackle this problem. Even the economic imperative that dominated much of our discussions was paradoxical. In a way, the economic imperative is a potent force that drives human action and so, this can be mobilised in the design of incentive and penalty schemes to influence human behaviour, provided human agency is well understood. At the same time, however, the dominance of economic framing of the problem can act as a major constraint to delivering on the sustainable development agenda. The obsession with performance measurement based on a narrow focus on quantitative targets of outputs and relatively less emphasis on the qualitative outcomes on the general well being of society, can result in a lack of holism in the way the sustainable development agenda is being approached.

Consequently, our interviewees acknowledged a lack of joined-up action across the diverse landscape of stakeholders involved. There is much that the government can do in coordinating responses from corporations and communities, through the role it plays as major procurer and regulator of the activities undertaken by the industry. Of particular interest to our interviewees are the need for greater clarity and transparency of policy intents around the sustainable development agenda, and a need for policy-makers to get closer to understanding the idiosyncratic nature of construction. Of course, the industry also can do much to ward off abdication of responsibility in this area. Our leaders insisted that practitioners can contribute to the contemporary agenda of sustainable development by remaining reflective and self-critical of current practices, to seek ways in which efficiency gains can be attained, and to re-invest such savings on basic research to advance our understanding of the key problems that face the industry and society at large. Furthermore, the modus operandi of focusing on short termism and mere subsistence is insufficient to steer the industry towards a sustainable future. What corporations need to do is to consider some of the longer term impacts of their present-day actions. And this can be achieved by encouraging diversity through the education system, in terms of integrating fresher perspectives from the younger generation and non-traditional sources of recruitment, as well as combining knowledge from a variety of disciplinary traditions.

Our interviewees agreed that the industry cannot act alone in resolving some of the critical issues associated with the sustainable development agenda. Engagement with policy-makers and the education system would be necessary if continuous improvement were to happen. Specifically, there were opportunities mentioned regarding the strengthening of connections between industry and academia. This is especially critical where research is concerned, as there is much to be gained in terms of getting practitioners and academic researchers to co-produce research outputs that could shed light on how to better manage construction activities and pursue technological advancements. Moreover, as we shall see later in this chapter, the role that basic research plays is critical in delivering the sustainable development agenda as there remains a lot of uncertainty in what constitutes knowledge about the agenda. Besides, research that is fed back into the education system would be instrumental in developing practitioners for a more sustainable future.

It is interesting to observe that when thinking about the future of the industry, our interviewees invariably talk about economic, social and environmental concerns, reiterating the rhetorical triple-bottom-line approach of sustainability. We have illustrated the key themes from our participants in Figure 3.1. For the rest of the chapter, we will review the state-of-the-art of the theoretical literature on sustainable development. In doing so, we set out to compare and contrast what our leaders think in relation to theoretical perspectives, so that overlaps and gaps in theory and practice for a sustainable future can be established.

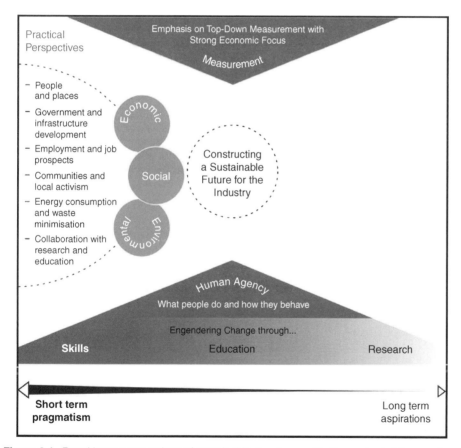

Figure 3.1 Practitioner perspectives of sustainable development in the context of the construction industry.

Sustainability: definitions and perspectives

The Pearce report[6] introduced the concept of sustainability by drawing on the UK Government's definition of 'ensuring a better quality of life for everyone, now and for generations to come'.[7] Despite its simplicity, the interpretation and adoption of sustainability principles are far from straightforward. Pearce later suggested that 'like all value-laden phrases, the seductive nature of sustainability has meant that different people define the term differently, partly in the anticipation that their own view of a good society will be accepted by others'.[8] Pearce's[9] contribution, as highlighted by Turner,[10] was to proffer a framework for understanding the application of sustainability in the British construction industry. This framework expands on the triple-bottom-line – economic, social and environmental perspectives – that has been identified by our leading figures above, by locating the discussion within four capital perspectives of man-made capital, human capital, natural capital and social capital. Accordingly, man-made capital perspective is concerned mainly with how the productive capacity of societies can be sustained, whereas human capital focuses on the long-term development of skills in the workforce. Natural capital looks at the how the planet's ecology can be preserved and enhanced over time and social capital focuses on the notion of communities.

At first glance, the issues analysed from the interviews in the preceding section correspond neatly with the framework adapted from Pearce.[11] The purpose of comparing our interviewees' perspectives and the theoretical literature is twofold. First, it is intended that the review will present the current state of knowledge, which perhaps explains the shaping of our leaders' thoughts about sustainability. Second, and probably more critically, the comparison will also reveal potential gaps in theory and practice of the sustainable development agenda, which might be useful for shaping the trajectory of how sustainable development could be addressed by the construction industry in the future.

Man-made capital: problems with an output-driven model

The construction industry is often credited as being the sector that underpins business, industry and general way of life in any country. In the UK, recent statistics from the Department of Business, Innovation and Skills (BIS) indicate that the industry employs around 1.26 million workers producing an output of just under £110 billion.[12] Though sizeable, Pearce[13] argued that official statistics tend to diminish the true picture of the industry as the data are derived from a narrow definition, which only accounts for the contribution of registered contractors. He estimated that broadening the definition of the industry to include inter alia the quarrying of raw materials, production and sale of building materials and products, professional services, self-build and the informal economy could imply a total employment of around 3 million people, contributing approximately 10% of UK GDP and around 70% of man-made capital (or manufactured wealth). Sustainable development, according to man-made capital perspective is largely about enhancing the productive capacity of the economy, often gauged by measuring productivity levels, i.e. how much goods and services are produced per input, and comparing these across industry sectors and/or across countries. Comparisons of productivity levels are often employed by governments to shape economic policy, especially in terms of steering towards which future economic activities to invest. In this section, we will discuss the difficulties of relying on productivity comparisons and question the reliability of measurement techniques and theoretical models that help explain productivity levels. Consequently, doubts are being cast on the efficacy of the narrow utilisation of productivity levels to drive economic policy for a sustainable future.

Difficulties of productivity comparisons

Measured through productivity levels, Pearce[14] observed that the industry's contribution to man-made capital, particularly in terms of infrastructure provision, lags behind our major competitors in Europe on a per capita basis. Elsewhere, there is also general recognition of the current housing shortfall to meet the demographic changes highlighted in Chapter 1.[15] These comparisons point to the long-standing productivity problem in construction. For example, Olomolaiye and colleagues[16] lamented about the suboptimal productivity of UK construction relative to the manufacturing sector. Others[17] have also suggested that the construction industry in the UK underperforms in comparison with such other countries as the USA. These trends seem to corroborate well with the portrayal of the UK productivity gap.[18]

Yet, there is also contrasting evidence to suggest otherwise. Barrett,[19] for instance, tracked the gross value added figures for the UK construction, manufacturing and service industries since 1975 and found that the construction industry actually outperforms manufacturing. Even the UK productivity gap with that of our European counterparts in official statistics appears to be closing. Recent evidence suggests that although UK construction is less productive than US construction, the gap has narrowed when compared to Germany and France. Ive and colleagues[20] estimated that the average labour productivity per person-year remains on a par with Germany. Similarly, when UK construction productivity is measured using value added per person employed, it was found that the UK's performance actually surpassed French and German construction productivity in 2001; thereby leading to policy recommendations that would see greater alignment with the pursuit of the 'best practice' agenda framed in the US construction industry rather than the development trajectories of the European construction industry.[21] Indeed, UK government policy currently advocates the US model as exemplary for closing the UK productivity gap and numerous comparative studies have been undertaken to attempt to explain the differences between UK and US productivity levels.[22] Researchers examining the retail sector, for instance, have argued that large UK businesses remain world class and are comparable to US firms in terms of productivity, suggesting that it is the 'long-tail' of smaller firms that requires catching up.[23] This observation is potentially interesting for the construction sector, which is dominated by smaller firms.

Limitations of using productivity comparisons to formulate economic policy

However, the formulation of economic policy based on international comparisons of productivity necessitates a degree of caution. As Griffith and Harmgart warn, '[...]

comparing productivity across countries can be deceptive [...] international comparisons can help us to learn about whether markets are working well, and about what sorts of policies are in place which are and are not effective [...] however, such comparisons are not perfect and should be interpreted judiciously'.[24] Broadberry and O'Mahony traced the historical context of international comparisons of productivity and argued that countries used for benchmarking against UK productivity fall in and out of fashion, and that competitor countries like Germany had once lagged behind UK in terms of productivity. They maintained that the current conviction for UK policy-makers to use the US model as exemplar may be myopic as there are limitations to wholesale copying of practices from other countries due to barriers such as geographic differences.[25] Specific to construction, Blake and colleagues noted that US construction differs from UK construction in that the US industry is dominated by larger companies, whereas the converse is true for the British industry.[26] Contextualising practices is, therefore, necessary for any policy adoption regarding productivity improvements.

Another area in need of further review is the adoption of information technology (IT) as a means of improving productivity. For instance, the transfer of this US policy to British industry remains very popular among UK policy-makers.[27] Yet, the construction industry globally – including US construction – has been consistently poor in the uptake of IT.[28] This is perhaps due to the industry's perception that the role of IT is somewhat limited when it comes to undertaking its core activity – the physical construction of buildings. Moreover, the UK construction industry is particularly dominated by SMEs that are less likely to have the capacity to invest in IT infrastructure. Commentators have argued for a rethink of the way IT adoption in construction companies, particularly SMEs, is measured and promoted; thus, reinforcing the need to make adjustments when embracing practices from abroad.[29]

Questioning the reliability of theoretical models and productivity measurements

Furthermore, the measurement of productivity is not without problems. Comparisons of productivity levels across countries are faced with the question of reliability, as there are often difficulties associated with measuring like-for-like because of differences in such market characteristics as pricing strategies and exchange rates.[30] Notwithstanding, critics have also observed that construction productivity measurements have not hitherto painted an accurate picture of the complexity of the production process. Crawford and Vogl,[31] for instance, suggested that the dominant emphasis on macro-level measures of productivity fails to account for the entirety of the construction process at the micro-level. This supports Chan and Kaka,[32] who noted that numerous studies undertaken to ascertain productivity at the construction project level merely focus narrowly on such stable activities as concrete operations. Radosavljevic and Horner[33] found that the reliance on statistical principles of normality in virtually all productivity measurements failed to deliver a better understanding of the ubiquitous complexity of construction productivity.

Even if productivity measurements were to be taken at face value, there also exists a fundamental problem with the basic assumption of existing productivity models applied to construction. Underpinning such models as the work–time model[34] and the CALIBRE approach[35] is the belief that working time can be dichotomised into productive time and unproductive time, and the general assertion that reduction of the latter results in a rise in productive time, and hence overall output. However, this binary relationship between productive and unproductive time has been proven to be non-existent in empirical data.[36] Indeed, using bricklaying as an example, Calvert and colleagues[37] remained baffled as to why global construction have not come to see even half of the production rates attained at the turn of the twentieth century when Frank Gilbreth's motion studies created a standard of 2700 bricks laid per bricklayer day.

Therefore, the focus on output that has become the principal driver for productivity measurements and comparisons offers only a partial understanding to our quest for sustaining higher productivity. Flanagan and colleagues[38] even suggested that it is perhaps time to move away from measuring outputs to the measurement of value derived from such issues as quality, innovation and organisational learning – what neo-classical economists like Robert Solow termed residual productivity – as they called for a paradigmatic shift towards understanding construction competitiveness rather than productivity. Indeed, earlier scholars have alluded to the tensions between two critical dimensions of the productivity problem: the first being that productivity improvements should yield an increase in output, an agenda that has come to prevail in managerial discourse; and the second being that productivity improvements should free up workers' time for more leisure.[39] However, the conventional dominance of emphasising the importance of increasing output has consequently led to a relative neglect on the perspective of the worker. As Chan and Kaka surmised, 'the constant fixation on the quantifiable aspects of productivity results in a regrettable dearth of attention to the labour element of construction labour productivity'.[40]

So, it seems that to secure a sustainable economic future, there are limitations on relying solely on output-based models of defining productivity levels. These measurements can often portray a far too simplistic picture that disregards the complexities of the production process (certainly in the case of construction), and the heterogeneous landscape of stakeholders operating in the industry. Furthermore, the measurement and comparison of productivity levels themselves do not guarantee that appropriate courses of action are taken to enhance man-made capital perspective of sustainable development. In framing the economic perspective of sustainable development, there is contemporarily more emphasis on sustaining the wider competitiveness agenda. This brings into question a range of other factors beyond the simplistic notion of productivity levels. In examining what Solow called the residual productivity problem, it is necessary to place more emphasis on enhancing the nature of resources over time as well. It is here that human capital perspective potentially offers some useful insights, which will be discussed in the next section.

Human capital: the rhetoric versus reality of investing in people

The concept of human capital has gained currency in recent times. In the UK, the Chartered Institute of Personnel and Development (CIPD) rallied the government to make businesses accountable for their human capital,[41] which led to the formation of the Accounting for People Task Force.[42] The concept of human capital is not new. Ever since Adam Smith examined the pin manufacturing process in the eighteenth century, the importance of human capital in the form of skills and dexterity of the workforce has been acknowledged. For Adam Smith, it was reasonable to see how enhancing the skills of workers alongside the division of labour would increase the productive capacity of the firm and worker. Indeed, Becker's theory of human capital[43] promises overwhelming support that investment in education and training generates rising personal income. Evidence can be sought from a greater proportion of higher level skills and IT adoption in US businesses and the associated productivity levels in comparison with the UK (see preceding section); and advocates of the importance of human capital would point to the successes of the Asian Tiger economies in the 1980s.

Yet, as Grugulis points out, 'the longstanding consensus that skills are "good things" [. . .] and the evidence that they can advantage every participant in the employment relationship have not been matched by a widespread adoption of high skills routes to competitiveness. Despite the existence of exemplary practice, extensive exhortations and official interventions, most jobs in Britain demand few skills'.[44] Even workers with degree qualifications are not guaranteed that their higher level skills would be put to good use at the workplace with the growth of GRINGOs (graduates in non-graduate occupations).[45] It is true, in fact, that the extant literature on construction skills paints a rather grim picture on the state of human capital investment.[46]

In this section, we examine the human capital perspective of sustainable development and delve into the reasons as to why disconnections exist between the rhetoric and reality of the skills development agenda. In part, the nature and structure of the industry characterised by the dominance of SMEs and reliance on casualised, flexible labour accounts for much of the lacklustre approach towards skills development. Yet, more fundamentally, the complexity of what skills really mean, coupled with the way public policy tends to simplify these in terms of quantifiable categories and levels, do little to instil confidence among employers to embrace the skills development agenda. The limitations of the economic approach to defining skills and expecting skills development to rationally happen are raised within this section. What is needed, again, is a holistic approach that accounts for the benefits of human capital that is accrued to workers and society at large, and this necessitates strong institutional coordination that encourages genuine collaborations and dialogue between the social partners of the state, employer and employee representation.

So why does the reality of the skills agenda not match the rhetoric of human capital theory?

If human capital is espoused to bring benefits of productive capacity for the firm and the worker,[47] why is there a mismatch between the rhetoric and reality of training and education investment? Construction researchers have offered a number of explanations for this phenomenon. The central argument tends to revolve around the nature of the industry. As mentioned previously, the construction industry is often used as a barometer of economic performance, and therefore is exposed to the vulnerabilities of economic cycles of boom and bust. Given such uncertainties, firms are less likely to engage in skills training and development, which requires a longer term view.[48] Furthermore, the industry epitomises the pinnacle of flexible organisation and its deeply entrenched reliance on self-employment[49] and contingent labour[50] reduces the industry's propensity to train. Together with the difficulty in attracting new entrants into the industry as a result of its beleaguering public image, it is inevitable that the industry consistently reports a low average training investment of around one-person-day per year.[51]

Generic versus specific skills: confusion over definition of skills

Ironically, the very definition of human capital used by economists potentially results in the lacklustre approach taken by firms in skills development. Becker's[52] notion of human capital is a simplistic one, which assumes a sense of rationalism. Mainstream economists tend to argue that one's sole purpose of survival is to maximise economic utility,[53] which when translated to skills development means the maximisation of one's productive capacity in order to maximise one's wage. Put simply, it is perfectly reasonable that one should invest in skills enhancement because one desires to become more productive. The concept of skills, under this narrow and rational economic perspective of human capital, implies that skills reside in the workers themselves as a measurably commodity so that any increase in skills (often measured through quantifiable levels) brings about a correspond-ing rise in worker productivity and wages – the concept of wage labour. Following from this, there is the question of who should then be responsible for investment of human capital. Becker's work consequently led to the distinction between firm-specific skills and skills that are completely general.[54] Thus, the investment of firm-specific skills should intuitively lie within the remit of firms, and the state, through its education system, should provide for the general skills as an increase in worker productivity should, in theory, benefit the economy as a whole.

However, commentators have argued that this dichotomy is far too simplistic. Groen commented that although 'this distinction is valuable at a conceptual level, it fails to describe the full range of training empirically. Most skills learned on the job may be somewhere between the extremes of firm-specific and general.'[55] He showed through an empirical analysis of the US labour market that as a market expands and becomes more competitive, employers tend to shift towards demanding more generic skills. So, the clear distinction between specific and general skills, and the accompanying consequences on determining where employer responsibility for training lies, becomes somewhat fuzzier; such a distinction does little to explain the reality of training that results in skills transferable to other firms.[56] Indeed, Groen criticised the relevance of Becker's model to reflect the impetus of the skills agenda in modern times, because human capital theory has hitherto assumed a '[. . .] model of investment in general and firm-specific skills, in which the training firm and the worker share the full amount of the return'.[57]

Yet, the concept of skills is indeed multi-faceted and boundaries between what benefits the employer or employee are not always distinguishable. Citing Cockburn,[58] Grugulis proffered three perspectives of skill: 'there is the skill that resides in the man himself, accumulated over time, each new experience adding something to a total ability. There is the skill demanded by the job, which may or may not match the skill in the worker. And there is the political definition of skill: that which a group of workers or a trade union can successfully defend against the challenge of employers and of other groups of workers'.[59] Accordingly, the first category relates to the conventional economic perspective of human capital where skills are commoditised for the maximisation of economic utility, and the latter two categories broaden to consider sociological implications of skills development on the worker and society respectively.[60] Indeed, skills could mean very different things to the worker and the employer, as Clarke observed in the construction labour process, 'whilst training creates skills, these skills have different values for the worker who owns, sells, employs and attempts to conserve them than for the builder *(employer)* who buys and consumes them [. . .] Under a capitalist mode of production [. . .] the determination of training provision is only possible through an analysis of changes in production and in the social relations regulating the labour process,'[61] thereby integrating the perspectives of skills at work and the socially constructed nature of skills definition between the actors concerned. However, as traditional configurations of what skills mean in terms of discrete categories and levels do little to reflect the complexity of what employers, workers and society demand of skills – if ever these can be clearly defined – the confusion that arises serves to impede engagement of the skills development agenda.[62]

Implications of complexity of skills definition on enactment of human capital theory

So, what are the implications of the expanded definition of skills? For one, the economic perspective promoted by human capital economists like Becker[63] is limited in attaining the ideal state of more skills leading on to greater productive capacity for both the firm and the worker. Instead, Groen[64] has demonstrated empirically the tendency for employers to pursue a strategy of more general skills and less specific skills as a market matures and becomes more competitive. Hence, employers are likely to abdicate from the responsibility of investing in skills development for fear that other employers would poach the workers once they have completed the training,[65] thereby resulting in continued de-skilling[66] of firm-specific skills. Recent analysis from the UK Labour Force Survey also provided evidence that skills-specificity in construction has declined over time.[67]

Grugulis and colleagues[68] suggested that this emphasis on generic skills offers, on the one hand, a false sense of upskilling among the workers, and, on the other, virtually no benefit of wage premium. Becker's[69] belief that investment in human capital would reap benefits of greater productive capacity and wage growth would stand to be tested, but this is certainly not the case in the UK construction industry.[70] For Grugulis, 'the key issue here is not that technology, market forces or flexibility do not support skills development: it is that they do not inevitably do so. There are choices to be made about the ways that work is designed, monitored and controlled, and these choices will affect skills in a range of ways'.[71] Arguably, the economic perspective of human capital emphasises human resources as an economic factor of production and potentially plays down the human benefits that can be accrued through development.[72] This has been recognised in the literature as the balance between hard and soft human resources management, where the former focuses on production and the latter acknowledges the welfare of workers.[73]

Grugulis and colleagues added that the concentration on generic skills meant moving 'the focus of attention away from the workplace and those who manage it, onto schools, colleges and universities, all of which have failed, it is alleged, to have imbued their students with the appropriate skills',[74] thus 'outsourcing' the responsibility of failing to achieve high-commitment, high-performance knowledge economy away from the realm of management. Insofar as skills matter, they lamented, 'they are not necessarily the first factor to be considered for the melioration of work, employment and the economy'.[75] Indeed, Beckingsdale and Dulaimi[76] observed that skills training and development is rarely seen as a core business activity among construction companies. Researchers also support this observation as they note that companies tend to approach such initiatives as Investors in People as a badge-collecting public relations exercise.[77]

The limitations of measuring skills: what else must be done?

When discussing man-made capital perspective in the preceding subsection, we identified the problems of formulating economic policy on the basis of measurement alone. So, would more measurement and reporting of human capital management improve the situation? Our best guess is probably not. As Elias and Scarbrough aptly point out, what is needed is 'evidence on the practice rather than on the theory of human capital evaluation', as they concluded, 'firms, while drawing on ideas such as the inclusion of people metrics in the balanced scorecard, actually adapted these approaches to their own company needs, preferring flexibility to standardised systems of reporting'.[78] These conclusions offer an explanation as to why standard reporting of human capital management remains elusive. Indeed, the perpetuation of standardised systems of measuring skills through levels of qualifications and the absence of evidence that this brings about upskilling is well documented.[79] So, measurement itself does not necessarily lead to improvements in human capital per se; what is crucial now is for firms and societies to start valuing labour.[80]

The view on human capital seems to be divided into two camps: economists who are largely concerned with economic performance and productivity, and sociologists who defend the power of labour in the social construction of skills. A compromise is perhaps needed to move forward in engendering real action from the debates. Sympathetic commentators have suggested that construction companies have to juggle between the short-term need for profitability and the long-term employee interests of skills development. Raidén and Dainty used the phrase 'chaordic organisation' to describe how construction companies deal constantly with both the chaotic business environment and the orderly, strategic planning of skills.[81] Clarke and Winch[82] called for a shift in educational philosophy in construction to include strong theoretical underpinning in schools, colleges and universities, work experience provided by employers and opportunities for simulation of work processes. This necessitates a deeper partnership between the education system and industry, and it is precisely this partnership that Leitch[83] felt his proposal for employment and skills boards could facilitate in moving the UK economy to a higher skills level.

However, institutions need to be strengthened so that genuine dialogue between the state, employers and workers can be maintained to ensure that appropriate skills are being developed for a sustainable future. In the UK, Broadberry and O'Mahony[84] argued that institutional structures have weakened since the Second World War, which resulted in the erosion of much-needed intermediate skills. They suggested following the lead of German institutions, which remained strongly supportive of maintaining a healthy balance between intermediate and higher level skills, a view shared by many others. Clarke and Herrmann[85] showed how differences in institutional structures between the UK and

Germany accounted for a more productive German construction industry. Indeed, institutions matter! And a clear example lies in the regulation of health and safety in UK construction – British construction remains one of the safest in Europe resulting from tightening up on legislation and there is evidence of a corresponding emphasis on health and safety training.[86] Yet, the perpetuation of neo-liberal economic policy and increased organisational flexibility[87] could potentially threaten the strength of institutions even in Germany.[88] We will examine the role of institutional structures in greater detail in the next chapter.

So, it appears that, theoretically, human capital perspective manifest in the skills development agenda is once again dominated by the rational economic approach, which, in turn, treats skills as quantifiable commodities in the form of categories and levels framed by public policy-makers. Yet, such an approach does little to translate into actual engagement of employers and employees in developing skills for a sustainable future. The notion that skills are a good thing, defined narrowly in terms of productive capacity and wage levels, seem questionable. Instead, there is mounting evidence that employers simply opt for development of generic skills, at the expense of de-skilling the workforce. Benefits of the human capital perspective of sustainable development accrued to workers and the wider society often escape public discussion. What is essential is not more measurement of fictitious skills categories and levels, but a concerted effort to coordinate an institutional response that encourages genuine engagement between the social partners of the state, employer and employee representation to safeguard skills for a sustainable future.

Natural capital: consensus gained or paradise lost

The green agenda has come of age. Not a day goes by without the mention of the C-word in popular media: our carbon footprint. From the way we commute, to the food we eat, to the houses we live in, the conscience of the British public is often pricked by the constant reminder that a radical lifestyle change is necessary to ensure the sustainability of the global natural environment. In 2006, a documentary film featuring former US vice-president Al Gore, *An inconvenient truth: a global warning*,[89] utilised powerful imagery to forecast potential scenarios of climate change, ranging from desertification to the dawn of a new ice age. In the film, Al Gore demanded immediate action from politicians, businesses and individuals as he maintained that there is not a single more important issue that warrants such compelling consensus from over 2500 scientists around the world. That consensus points to the grave situation that business-as-usual would result in the depletion of Earth's natural resources, which, in turn, would threaten the very existence of humankind. As this section unfolds, we present the key arguments on natural capital that suggest that scientists are far from reaching consensus, especially in terms of mitigation responses. We examine the main debates in the study of natural capital and suggest that, as in man-made and human capital, the challenge lies in moving from the rhetoric of measurement to the reality of action.

A consistent theme emerging from a review of man-made and human capital perspectives of sustainable development is the limitation of framing the agenda in purely quantifiable, economic terms. We see this again in the literature on natural capital perspective, where proponents of securing a more sustainable future for the planet's ecology are compelled to consider environmental problems as economic ones in order to demand the attention this deserves. Yet, it is observed that such an approach drives policy-makers to be fixed on measurement alone and does little to promote actions of a strong environmental sustainability nature, i.e. one that not only ensures a minimisation of harm to the natural environment, but also encourages the preservation and replenishment of natural stocks. In this section, we discuss the limitations of the economic approach in greater depth, and explain why a strong environmental sustainability approach remains elusive. Ostensibly this is because knowledge about the workings of natural capital remains incomplete and there are problems associated with assigning monetary value on natural stocks, especially where the longer term value for the benefit of future generations is concerned. We also discuss how incomplete knowledge results in less prescription of public policy in this area; a corollary of such vagaries is that policy implementation to ensure strong environmental sustainability remains ineffectual.

The need to frame the environmental agenda in economic terms

The concept of natural capital pertaining to the sustainability agenda was popularised in the 1980s by the late Professor David Pearce.[90] According to Turner, Pearce viewed natural capital 'as a fruitful way of integrating ecological sensitivities into mainstream economics'.[91] Admittedly, it can be observed that the environmental agenda is now given a more prominent place on the global stage as political leaders from across the world continue to debate both the implications of climate change on the economy, and, more critically, effective mechanisms to mitigate the cost of meeting the environmental sustainability agenda. Reviewing the Stern report,[92] Shipworth observed that Stern's recommendations are 'now resonating around Europe and to a lesser extent America and shifting the debate away from the science of climate change to the economics of action versus inaction',[93] adding 'while there is little that is surprising in these recommendations, the overall theme is clear. Most economists favour market mechanisms where available, and regulation where there are market failures'.[94] As we shall see later, market failures do exist when adopting strategies to mitigate the impacts of climate change. Deciding on regulatory responses, however, remains contentious.

It is fair to say that few would dispute climate change to be an issue of the day. Particularly in the West where governments struggle to encourage the electorate to vote, today's politically apathetic youth are likely to be interested in such single issues as climate change and the environment. Yet, believing that good citizenry – or free market mechanisms in economic speak – is all that is needed to tackle such issues is simply naïve. Pearce and Atkinson commented that 'the concept of "natural capital" does not adequately conceptualise the economy–environment linkage. Only a comprehensive "ecological economics" can do that and we do not believe that a coherent body of thought has yet emerged [...] we argue that the forces bringing them together are most likely to emerge by forcing existing paradigms to account for environmental problems'.[95] What is interesting is the element of compulsion that they placed on considering environmental problems as part of economic ones.

Distinction between strong and weak sustainability

Researchers from the field of ecological economics have distinguished between two forms of sustainability: weak sustainability and strong sustainability. Devkota,[96] drawing on inter alia Pearce and Barbier,[97] provides a concise explanation of the distinction between weak and strong sustainability. The weak sustainability perspective views natural capital and man-made capital as substitutes, such that the depletion of any one form of capital is acceptable insofar as reinvestment in other forms of capital takes place in order to maintain or increase the total stock of natural and man-made capital. The measurement of weak sustainability tends to follow a cost–benefit approach and is invariably commensurable with monetary valuation. The strong sustainability perspective, on the other hand, considers natural and man-made capital to be complementary and that all natural and other forms of capital should not only be kept independently intact over time, there is also a need to maintain essential, non-replaceable and non-substitutable environmental resources.[98] Therefore, the preservation of the natural environment and its resources is central under strong sustainability considerations. Shipworth states that strong sustainability 'evokes the precautionary principle [...] a legal mechanism for managing the variability in impact arising from one's uncertainty of the consequences of actions considered likely to cause serious or irreversible environmental damage'.[99] However, Turner[100] argued that green national accounting measures like 'Genuine Savings Measure', 'Index of Sustainable Economic Welfare' and the 'Genuine Progress Indicator' have demonstrated that many countries cannot even pass the weak sustainability test, let alone provide evidence of attaining strong sustainability.

Why is strong sustainability elusive? The problems of incomplete knowledge

Neumayer suggested that the pursuit of environmental sustainability, especially strong environmental sustainability, can be a struggle: 'If the current generation still thinks that additional precautionary action is warranted, it should do so [...] Natural and economic science is able to guide in making this decision transparent and rational. It will not be able to give the answer in the society's stead, however [...] both the natural and economic science of global warming is unable to provide unambiguous answers about how much emission abatement is warranted. Uncertainty and ignorance are too widespread'.[101] So, what is suggested here is that scientific knowledge about the agenda remains incomplete and that present ignorance about the subject means that any intervention to address the environmental agenda will invariably entail some form of moral consideration.

One key area that still remains debatable about the environmental agenda is what actually constitutes critical natural capital. Ekins suggested that when understanding natural capital, one needs to consider the functions of natural capital for human beings and the functions of natural capital itself.[102] He argued that whereas we have a better understanding of the impacts of natural capital on human beings (i.e. 'functions for') in such terms as the economy, waste management and human health, our understanding of the 'functions of' natural capital remains elusive. For Ekins, 'the "functions of" the environment are those which maintain the basic integrity of natural systems in general and ecosystems in particular. These functions are not easily perceived, and scientific knowledge about them is still uncertain and incomplete. What may be said with certainty, however, is that whether science understands these functions or not, and whether people value or are ignorant about them or not, the continued operation of the "functions of" the environment is a prerequisite for the continued performance of many of the "functions for" humans'.[103]

In other words, the natural environment is taken as a good thing in the abstract, but just how this natural environment operates is still being fathomed by humankind. As a consequence, any scientific knowledge and intervention that arises from this knowledge will always remain partial. Yet, he urged for more research to comprehend the dynamic complexities of the functions of the natural environment, as when 'looked at in isolation, these "functions of" the environment may appear useless in human terms, and therefore, dispensable. Considered as part of a complex natural system, these functions may be essential for the continued operation of other functions of much more obvious importance to humans. The danger is that the isolated view, or scientific ignorance about the natural complexity, may result in "functions of" being sacrificed for economic or social benefits, without appreciation of the wider implications'.[104]

Ekins maintained that 'there is little current evidence that societies are actually prepared to put it into practice'.[105] Several commentators, including Pearce himself, remain sceptical about the precautionary principle because of the ambiguity that surrounds it.[106] Furthermore, there exists an unresolved, hotly debated issue of how far into the future the current generation should account for.[107] Regarding preservation of the environment, this brings in the subjective concept of aesthetics. We will visit this issue when we discuss social capital below, but it is interesting to note that Pearce[108] himself avoided affording an estimate on this very issue when he was examining the sustainability of the UK construction industry.

Difficulties of measuring natural capital

Measurement, especially in the economics literature, necessitates a high degree of quantification, and those issues that cannot be easily quantifiable appear to be sidelined. This is potentially suicidal[109] when adopted in public policy. In fact, England suggested that measurement of natural capital can be a waste of intellectual effort as he 'reached the conclusion that, although natural capital is a powerful metaphor worthy of retention by ecological economists, its precise measurement should not be at the top of our collective research agenda'.[110] Of course, what measurement fails to achieve is to shed light on what mitigation responses are required, or even what response is effective or not. Like many accounting measures, the indicators are also at best lagging behind practice. As Wackernagel and colleagues commented 'Present demand that damages future supply will only show up in future Footprint assessments. The Footprint and Biocapacity thus derive directly from prevailing yields, and do not make adjustments for "good" or "bad" management practices'.[111]

Furthermore, Chiesura and de Groot observed that any measurement approach that simply represents in monetary terms serves only to perpetuate individualism and ignores the plurality of the human experience in the forms of cultural and non-material well being.[112] They advocate a socio-cultural approach to define natural capital as they asserted: 'Central focus of the socio-cultural perspective is thus the human being with its social and psychological context, its non-materialistic needs, its understanding of well being, and the rational as well as the emotional components of its attitudes towards the natural environment'.[113] Indeed, human well being needs to be re-evaluated beyond what the mere economic cost–benefit analysis can examine.[114] Here again, Pearce[115] recognised the growing interest of the existence, even the importance, of the 'happiness' literature,[116] but seemed quick to dismiss its relevance and credibility in the pursuit of sustainability.

The ethical dimension: dilemma between present energy needs and benefits for future generations

If measurement is a problem, then ethical consideration is even more contentious. As Neumayer noted, 'the answers are dependent on the underlying ethical decisions concerning how much to take the future welfare into account and whether one thinks that what future generations care about is only total capital or specific sub-categories like natural capital. Ultimately, it is on us to decide whether we think consumption growth can compensate future generations for damage to natural capital and human health or not'.[117] This returns to the paradox of dealing with short-term imperatives and long-term benefit discussed above.

Stern talks of bridging both intragenerational and intergenerational gaps in the quest of maximising utility in the economics of climate change.[118] In other words, there is a need to ensure the sustainability of the natural environment across time and space. However, this is a somewhat rose-tinted view of the reality of action (or inaction) on climate change. Wackernagel and colleagues asserted, 'In fact, those contributing most to climate change through their energy intensive lifestyles will most likely be less affected by, and better shielded from, the outfalls of climate change than poor people living on marginal land or in underserved urban conditions. Such disparities between those who profit from resource consumption and those who bear the environmental burden strongly encourage overuse of resources'.[119] This perhaps explains why the practice of carbon trading, where the developed world can in principle continue with business-as-usual by financially offsetting their high emissions against those of the less developed world, remains more favourable among businesses than say, climate change taxation; or perhaps why we have seen headline stories about how biofuel crop plantations (where the science remains unclear) are exacerbating poverty in developing countries.

Indeed, consumption particularly in the West is insatiable. Herring argued that restriction of consumer behaviour in terms of energy consumption has never been favourable by the public: 'This concept of sufficiency with its emphasis on reducing consumption and "living well on less" [. . .] so far remains a mainly ethical exhortation rather than practical approach to Western consumers. However, it is being developed by academics and encouraged by governments in a weak form under the slogan of "sustainable consumption"'.[120] Herring (2006) went on to conclude with an ethical question on what a 'good' life is, as he suggested, 'a consensus and practical solutions to devise a "conservation" or sufficiency lifestyle will take time. In the meantime, energy efficiency is a valuable tool to save consumers' money and stimulate economic productivity and it should still be promoted whatever be its impact on energy consumption'.[121] There is perhaps a tendency to feel that as the past is long gone and the future is uncertain, what remains secure is the present. Such an attitude would challenge any proposition of a lifestyle change among consumers. Indeed the extant literature promotes an understanding of the functions of natural capital for humans.

However, England suggested a change in perspective to look at how humanity fits into nature: 'we retain "natural capital" as a pedagogical device and as a component of our preanalytic vision of how humanity fits into nature [...] It follows that human society is approaching, or perhaps has already entered, an era when scarcity of natural capital constrains aggregate output'.[122] Paradoxically, consumption is to be restricted by the limits of natural capital. Thus, if consumption is to be safeguarded as a right of future generations, an ethical stance for action is now required. Stern[123] believes that it is inappropriate to consider the rights of future generations as intrinsically less valuable than those of the current generation.[124]

Thus, Stern's[125] proposals on adjusting discount rates to reflect concern for the future could signal a departure from conventional economics of deriving net present values. But, for the moment, individual self-interest is likely to prevail. Spash's conclusion about the Stern report is illuminating on the issue of ethical consideration: 'the approach taken clearly allows traditional economic growth to be defended. The argument avoids the fundamental question of why more consumption and production is necessary. Indeed to ask such a question is economic heresy because such growth is the foundation of modern political economy, where the consumer is mythically sovereign, firms have no political power and governments hardly exist. That this orthodox economic model might be failing and is impossible to sustain goes to the heart of ecological economics'.[126]

Specifically within construction there is the long-standing problem of influencing consumer behaviour to take into account whole life cycle assessments. According to Turner,[127] life cycle assessment represents one way of evaluating strong environmental sustainability. In theory, life cycle assessments allow users to develop a long-term view of costs beyond that of initial construction. Academics and practitioners have often cited 1:5:200 as the ratio of construction costs to maintenance and building operating costs to staffing and business operating costs. This ratio is significant because it illustrates the potential value of how the initial design and construction process can impact on the overall value of running the facility and operations over the life of the building.

However, recent analyses of empirical evidence have revealed that the value of operating costs is over-inflated. Hughes and colleagues examined published data of three commercial buildings in the UK and found that a more realistic ratio is 1:0.4:12.[128] Building on data on Central London offices, Ive revisited the base assumptions of the original ratio and made adjustments for such issues as discount rates, construction costs, land values and productivity and found that a more plausible ratio seems to be 1:1.5:15.[129] Ive went a step further by arguing that when developing such a ratio, one needs to distinguish between types of occupancy, e.g. commercial versus non-commercial, developer versus owner–occupier, etc. For him, 'it is therefore important that tenants, property investors and developers intending to sell tenanted offices to investors can all see the relevance of the argument about the relative size of the ratios to them'.[130] Accordingly, developers intending to sell or rent tenanted offices are more concerned with the market values they can get from sale/rental and less inclined to consider 'cradle-to-grave' options. Thus, this reinforces yet again the short-term view that is so prevalent in society, which can consequently hamper any effort towards strong environmental sustainability.

Problems with policy implementation

We assert that the critical issues of the measurement problem, ethical considerations and lack of long-term view create problems of clarity in defining policy instruments to address the environmental agenda, which, in turn, translates to difficulties in policy implementations. In the construction sector, this can be seen in the contemporary low-carbon agenda. The delivery of zero-carbon housing is one of the most recent aspirations of the UK government in the pursuit of greater energy efficiency and the Kyoto protocol targets.[131] The residential sector in the UK accounts for about 30% of the UK's total carbon emissions.[132] Boardman and colleagues observed, however, 'Contrary to experience in most countries, UK carbon emissions have fallen in recent years, being around a fifth lower in 2003 than in 1970'.[133] Therefore, although reduction in carbon emissions is vital, curtailing energy consumption is more important in the formulation of UK energy policy. After all, as Banfill and Peacock[134] argued, one of the major impetuses driving public policy and regulatory change in the UK is the security of energy sources to maintain projected energy-intensive lifestyles, rather than genuine engagement with the carbon agenda.[135]

Despite the boldness and laudability of the policy proposals, the reality appears once again detached away from rhetorical aspirations. Schiller suggested that debates surrounding the contribution of construction towards sustainable development had hitherto been emphatically framed around the aspects of new buildings.[136] He maintained that attention needs to be given to the provision of urban infrastructure, which is arguably as resource-intensive as new-built projects, if policy-making were to derive a long-term view. What Schiller highlighted is the incompleteness of knowledge surrounding the construction industry's contribution to sustainable development.[137]

There is the further issue of refurbishing and adapting the existing building stock. Boardman and colleagues concluded, as part of the 40% house project, that there needs to be an increase in the demolition rate to 80 000 dwellings per annum across the UK, a rate last achieved in 1975.[138] There are indeed concerns as to whether current industry's capacity can cope with such a scale of demolition.[139] Similarly, Banfill and Peacock,[140] when critiquing the policy on zero-carbon housing, suggested that both institutional mechanisms and the industry's supply chain were currently inadequate to meet the proposed targets by 2016. Lowe remained optimistically cautious: 'The conversion of the UK housebuilding industry and supply chain to one capable of delivering 160 000 to 200 000 passive houses per year by the middle of the next decade will be an enormous task. If the UK is ultimately successful, it will have achieved more in the next seven years than Germany, where the standard has been developed, has achieved in the last 17'.[141]

To succeed, there needs to be political urgency and a strong will for implementation to milestones in a set time-scale. In order to do this, there must be a degree of clarity, transparency and prescription in public policy and legislation so that real action by industry can be encouraged. However, far too often, the political shift towards arm's length approach by government implies that methods of implementation are not always laid out clearly. Recent experience with changes to Part L of the building regulations that govern energy efficiency of buildings[142] demonstrate this problem, and the success of political will remains to be seen because the vagaries of performance-based building regulations[143] means that there is little prescriptive guidance to ensure practitioners can act on the agenda effectively.

The challenge of altering human behaviour

And if the challenge of political will is not enough, there is the battle to change the hearts and minds of consumers. As mentioned previously, consumption growth has been the underlying assumption driving public policy. Arguably, this somewhat pessimistic approach is due to the fact that the knowledge on current consumer behaviour regarding the use of buildings from an energy perspective remains patchy.[144] Still, there is growth on work in this area. Wood and Newborough investigated how the use of domestic appliances can lead to potential savings in energy.[145] Pett and Guertler examined how people actually use energy-efficient systems installed in their homes.[146] And ongoing work at University College London (www.bartlett.ucl.ac.uk) should shed light on how occupant behaviour in relation to air-conditioning could impact on energy consumption. However, these studies captured a static snapshot of consumer behaviour through such techniques as surveys and interviews[147] and controlled experiments.[148] More research needs to be done to examine consumer behaviour from a holistic and dynamic approach, and, until such evidence is available, researchers can only rehearse the need for adjustments in taxation/incentives[149] and the education of consumers[150] at an abstract level.

To sum up, it is observed once again that there exists a gap between the aspirations of preserving and enhancing the natural environment and the reality of not meeting the strong environmental sustainability criteria. The discourse surrounding environmental sustainability is also dominated by the economic perspective. Paradoxically, the economic focus has enabled political leaders to develop some kind of a common language to facilitate discussions about addressing the environmental problem on a global scale. At the same time, the economic focus also means that often the actual agenda is about cutting the costs of energy consumption rather than more genuine concerns about climate change and the low-carbon agenda. A recurrent theme in the literature is the inherent tensions that arise in balancing human needs in the present and safeguarding the natural environment for future generations to come. As a result, inaction derived from a sense of conservatism is inevitable, given the incompleteness of knowledge about the subject. Meeting strong environmental sustainability necessitates a response far grander than that which can be garnered at the individual level, thus requiring yet again an institutionally coordinated approach where cooperation is sought between governments, corporations and communities to engender action. No intervention, however, can be made effective without understanding how human behaviour is shaped as society and technology use advances. At this point, it is appropriate that we now turn to the social dimension by reviewing the social capital perspective of sustainable development.

Social capital: building trust and sustainable communities

In a BBC documentary entitled *How We Built Britain*,[151] presenter David Dimbleby reveals how the built environment, through its architecture, can inspire people in their social and economic activities. In one episode, he recounts British philanthropy as he visits Saltaire, a Victorian industrial village north of Bradford in West Yorkshire, UK. Saltaire was designated as a world heritage site by UNESCO in 2001 (see www.saltaire.yorks.com). However, its humble beginnings stemmed from the vision of its founder, Sir Titus Salt, Mayor of Bradford in the 1850s and a businessman in the clothing industry. Titus Salt was aware of the acute poverty that struck the Victorian working classes, and so he embarked on a project to build, among other things, decent houses, a school, hospital and leisure facilities in an attempt to elevate living standards for his workers. The efforts were remarkable at a time when the struggles of the working classes were largely ignored by fervent capitalists, and, arguably, increasing worker welfare probably led to higher production in Salt's mills. It is perhaps unlikely that Titus Salt labelled his achievements in Saltaire as the creation of a sustainable community or indeed a contribution to social capital – but this is precisely what he did!

Notwithstanding Salt's intentions (whether altruistic or not), the development of places like Saltaire deserves greater examination, especially given current political interest in the sustainable communities' agenda. For Bridger and Luloff, this renewed interest in sustainable communities can be explained through the politics and economics of globalisation and urban growth: 'policy recommendations [. . .] and policy discussions [. . .] intended to reduce sprawl and create more livable communities. Some might argue that this shift can be explained as a manifestation of the [. . .] rhetoric of individual responsibility and devolution of responsibility to the state and local level [. . .] it is arguable that the renewed interest in the local community is one of many, often contradictory, responses to globalization and economic restructuring [. . .] on the one hand, communities find themselves in fierce competition to attract mobile capital [. . .] on the other hand, the political and economic processes that commodify and homogenise places have provoked growing resistance and sparked attempts to construct alternative conceptions of community life'.[152] Salt's primary focus appeared to hinge on three things: a roof over one's head, a place for work and leisure, and opportunities for maintaining (and developing) one's well being. A century and a half later, these basic principles still hold true for the development of sustainable communities.

But, far from being straightforward, the definition of sustainable communities is fraught with problems and is continually debated by policy-makers, practitioners, academics and communities themselves. In this section, we review some of these definitions of what a 'sustainable community' encompasses. Although there are only subtle differences between various definitions of the concept of 'sustainable community', the reality of 'doing' sustainable communities appears to be detached from the theoretical and policy aspirations. In part, this is because of the challenges of developing meaningful and effective engagement and participation from the grassroots within communities. Genuine involvement is often hindered by the top-down approach of conventional planning systems, where government and industry professionals establish the aspects that are deemed critical for sustaining communities. Yet, these criteria are driven largely by economic concerns, for instance, through the manifestation of efficiency gains in the design, construction and use of buildings and the reduction of transaction costs associated with contractual relationships. Once again, we stress that planning criteria are often motivated by policy-makers' desire to standardise measurement, and this deterministic process ignores the democratic process that potentially places value on more intangible aspects of human well being in the quest for attaining sustainable communities.

Definitions of sustainable communities

In theory, modern definitions of a sustainable community, and the aspirations that follow, do not differ much. Bridger and Luloff,[153] for example, suggested five dimensions of a sustainable community including the emphasis of increasing local economic diversity, the need for economic self-reliance, the reduction of energy use, the protection and en-hancement of biological and environmental diversity and the commitment to social justice. Drawing on the work of Girardet[154] and Lord Richard Rogers,[155] Doughty and Hammond pointed to the key perspectives of 'sustainable cities' in terms of beauty, accessibility, proximity, creativity, diversity, ecology and social justice.[156] Doughty and Hammond highlighted the political impatience with the concept of sustainable cities,[157] but observed that the UK government adopted these key principles when public policy shifted towards 'sustainable communities'. Indeed, the then-Office of the Deputy Prime Minister, in their manifesto to provide affordable *Homes For All*,[158] reiterated the importance of sustainable communities to provide active, inclusive and safe environments that are well-run, well-designed and built, well-connected, well-served, thriving, environ-mentally sensitive and fair for everyone. The striking similarities of these definitions are elaborated in Table 3.1,[159] which clearly shows how the sustainable communities' agenda cuts across economic, environmental and social themes. Arguably, apart from the contemporary focus on the environment (which incidentally invokes our earlier discus-sion on the complexities surrounding its measurement earlier), the pursuit of sustainable communities in today's terms bears much semblance with what philanthropists like Titus Salt set out to do over 150 years ago.

'Doing' sustainable communities: challenges of ensuring community participation

Yet, the reality of 'doing' sustainable communities presents more challenges when compared to 'thinking' in policy terms. Bridger and Luloff argued, 'In our opinion, the central theme behind many of the recent attempts to recapture a sense of community is the recognition that such a task requires alternative constructions of place – symbolic, economic, and physical constructions which reduce the alienation of people from one another and from the environment'.[160] It is precisely this reduction in alienation that the extant literature discusses the need to involve, and seek the buy-in, from the commu-nities concerned when developing such symbolic, economic and physical constructions of what a sustainable community looks like. However, as Bridger and Luloff point out, 'community studies document numerous barriers to broad involvement and the high level of activeness envisioned by proponents of sustainable community development'.[161] They added, 'leadership and participation are largely limited to local elites whose interest

Table 3.1 Definitions of 'sustainable community'

Emphasis	Bridger and Luloff (2001: 462–463)	Doughty and Hammond (2004: 1231)	ODPM (2005: 4)
Economic	"[. . .] there is an emphasis on increasing local economic diversity (p. 462)" "[. . .] virtually all definitions stress the importance of [. . .] the economic self-reliance [. . .] the creation of local markets, local production and processing of previously imported goods, greater cooperation among local economic entities (p. 462)"	"A Diverse City, where a broad range of overlapping activities create animation, inspiration and foster a vital public life" "A Creative City, where open mindedness and experimentation mobilise the full potential of its human resources and allows a fast response to change" "A Compact and Polycentric City, which protects the countryside, focuses, and integrates communities within neighbourhoods and maximises proximity"	"Thriving: with a flourishing and diverse local economy" "Active, inclusive and safe: fair, tolerant and cohesive with a strong local culture and other shared community activities" "Well run: with effective and inclusive participation, representation and leadership" "Well connected: with good transport services and communication linking people to jobs, schools, health and other services Well served: with public, private, community and voluntary services that are appropriate to people's needs and accessible to all"

Environmental	"[. . .] reduction in energy use coupled to the careful management and recycling of waste products (p. 462)" "Sustainable communities provide a balance between human needs and activities and those of other life forms (p. 463)."	"A City of Easy Contact and Mobility, where information is exchanged both face-to-face and electronically" "An Ecological City, which minimises its ecological impact, where landscape and the built form are balanced and where buildings and infrastructures are safe and resource efficient" "A Beautiful City, where art, architecture, and landscape spark the imagination and move the spirit"	"Well designed and built: featuring a quality built and natural environment" "Environmentally sensitive: providing places for people to live that are considerate of the environment"
Social justice	"Sustainable communities provide for the housing and employment needs of all residents and they do so without the kind of class and race-based spatial separation that is typical of many localities (p. 463)"	"A Just City, where justice, food, shelter, education, health, and hope are fairly distributed and where all people participate in government."	"Fair for everyone: including those in other communities, now and in the future."

in development often has much more to do with private gain than community well-being'.[162]

Indeed, there are limits to community participation and consensus-building among end-users in shaping sustainable communities. Evidence of good practice remains patchy, and small-scale examples on delivering schemes with user input tend to be confined to certain segments of public sector building.[163] In the UK, developmental work on design quality indicators[164] is intended to provide a common set of measurement criteria for stakeholders to engage in a dialogue about design of the built environment. It is perhaps appropriate that Whyte and Gann[165] entitled their paper *Design quality indicators: work in progress*, as several authors subscribe to the view that it is extremely difficult to evaluate design in an objective sense.[166] Exploratory work has tried to get stakeholders to self-assess such intangibles in the school environment as calmness and security and safety when making decisions across schemes.[167] In their conclusions, they discussed the degree of uncertainty with regards placing a value on such intangible measures. Furthermore, there is the question of who gets to vote on these measures from the community; the risk of participant selection is indeed problematic.[168] Macmillan asserted that more needs to be done by both researchers and professionals in terms of creating evidence-based design that integrates the valuation of intangibles.[169]

Until such measures are considered, it is always easier to measure the efficiency of a building function than it is to measure the concept of design. Similar to the discussion above on man-made, human and natural capitals, the practice of creating sustainable communities has a tendency to fall prey to the dominance of the economic perspective. It is here that understanding the historical context of the development of master planning in the UK can be useful. Giddings and Hopwood, for instance, traced this development since the 1950s and noted the tension between the desire to engage with community participants and the inherent expert status that is often guarded by built environment professionals: 'sustainability relates to people, building on existing and mutually supportive activities and encouraging residency, public places and spaces, help for the local economy and the concept of the city as a process. Whereas master planning is generally a top-down approach by experts, often clearing out existing activities, creating large single use areas of private or ambiguous ownership [. . .] The usual strategy is to have new large master planned complexes often for retail or leisure. However, the benefits of such actions are questionable, especially when compared with the many costs, both the money spent on their construction and the loss of small businesses, social networks and sense of place'.[170] Arguably, whereas the political rhetoric purports to encourage community participation in making decisions at the local level, there is still a preference for the familiar practice of top-down decision-making by policy-makers and professionals.

The limits of top-down engagement: trust and emancipation of the grassroots

Yet, trust for policy-makers and professionals remain increasingly dubious in modern times. A study undertaken in Saskatchewan, Canada found that respondents tended to place their trust on people from nearby communities the most, experts and professionals less strongly and governments least of all.[171] If this finding were generalisable, then the implications for the way sustainable communities are conceived and implemented by policy-makers would need to be examined. After all, as Coleman, who popularised the concept of social capital, suggested, 'Social capital [...] comes about through changes in the relations among persons that facilitate action [...] for it exists in the *relations* among persons. Just as physical capital and human capital facilitate productive activity, social capital does as well. [...] The value of the concept of social capital lies first in the fact that it identifies certain aspects of social structure by their functions [...] the value of these aspects of social structure to actors as resources that they can use to achieve their interests'.[172] Understanding trust relationships between the social actors (e.g. state, professionals, community users) would be of utmost importance in shaping and determining the shared interest of a sustainable community (if this were a shared aspiration to begin with).

Indeed, proponents of social capital theory suggests that building trust among social actors would lead to a decrease in reliance on bureaucratic contracts, which, in turn, would result in a reduction of transaction costs,[173] thereby bringing greater efficiency. This rationale is to provide the driving force for much of the partnering efforts seen in the construction industry today. Researchers have examined the nature of trust in construction. Swan and colleagues,[174] for example, investigated key dimensions of trust in construction, including honesty and openness, fairness and reasonableness, promise-keeping, mutuality and reciprocity, values and ethics and reputation. Through examining the social networks that exist in construction projects, they developed a trust inventory to determine the social structure of trust in construction. Taking a behavioural approach,[175] Smyth maintained that trust is an emotional concept as he went on to explain the key emotions of pride, confidence, fear and humility in relation to the construction project environment.[176] These studies are informative, especially at a time when there is much discussion about a cultural change in an industry crippled by a low-trust environment.[177]

However, the utopian view of greater trust leading to the abandonment of contracts and a decrease in economic transaction costs appears to be misguided in reality. Although the industry has seen a rise in partnering activity, its success is not as widespread as some were led to believe. Whereas there is growing interest in the field of relational contracting, the legal profession still earns a fortune from an industry that continually gets into dispute over contractual terms. In fact, we have not seen the disappearance of lengthy contracts promised by social capital theorists. Instead, recent evidence from the information systems industry suggests that complex contracts might just go hand in hand with greater preference for relational contracting. Through a survey of senior executives and managers, Poppo and Zenger found that 'managers tend to employ greater levels of relational norms as their contracts become increasingly customized, and to employ greater contractual complexity as they develop greater levels of relational governance'. This led them to conclude that there is 'a need to explore more carefully and predict more cautiously the relationship between formal contracts and relational governance'.[178]

Similar to our arguments made about the other capitals previously, the limitation of understanding sustainability from a social capital perspective is limited by economic dominance. Coleman asserted, 'the concept of social capital constitutes both an aid in accounting for different outcomes at the level of individual actors and an aid toward making the micro-to-macro transitions without elaborating the social structural details through which this occurs'.[179] This assumes, once again, that it is possible to measure objectively the different outcomes in a disaggregated fashion. Such an economic approach leads to the problem of reductionism, which consequently results only in partial successes in policy formulation and implementation. In our discussion on sustainable communities, we have shown that it is just as important to elaborate the social structural details, which is given scant attention by theorists like Coleman, who advocates rational action theory where 'people are viewed' as purposive agents who make rational, deliberate choices to maximize their utility'.[180]

The need to understand human behaviour is once more being emphasised, and so this requires a deeper sociological exposition of the relationship between the changing social structures and agents involved to fully appreciate the power of social capital in our quest for sustainable development. In so doing, it is essential to examine how decisions on sustainable communities are made in such a collective manner that transcends the utility of individual agents and departs from mere abstract notions of trust and altruism. There is again a need, therefore, for an institutionally coordinated response where the genuine dialogue between the state, corporations and community actors is being effectivised, and that appropriate representation and democracy are being protected in the shaping of sustainable communities.

The measurement problem: are efforts towards sustainable development doomed to fail?

We have attempted to sketch out the developments in the literature surrounding the key concepts of man-made, human, natural and social capitals in the pursuit of sustainable development based on a framework initiated by Pearce. The theoretical perspectives are illustrated in Figure 3.2. From this salient review, a number of observations can be made. First, there is a wealth of knowledge established on measuring the various capital perspectives of sustainable development. So, whether it is about cross-country comparisons of productivity, or skills levels depicted by human capital investment, or carbon emissions and energy consumption, or the levels of trust in social relationships, there is a huge preoccupation with measurement. However, measurement is only as good as the criteria chosen, and our discussion above also revealed an overwhelming lack of consensus regarding both the criteria and the final measurements themselves. Knowledge about the various capitals and their contribution to sustainable development remains, at best, incomplete.

We have argued that partial knowledge is due to the dominance of the economic approach in setting the sustainable development agenda across all four capital perspectives. Although the economic approach appears to arm policy-makers with a common language with which to discuss critical issues about the agenda, this brings about a number of limitations. First, economic measurements tend to rely largely on the ability to place a monetary value on some chosen criteria. The process of how such criteria are chosen, and by whom, is relatively less articulated in the literature, and there is a need, therefore, to research the unexplored terrain of power relations that govern such approaches.[181] Furthermore, we have repeatedly shown that this monetary evaluation alone can be arbitrary and lead to a level of reductionism that is often unhelpful for both policy formulation and implementation.

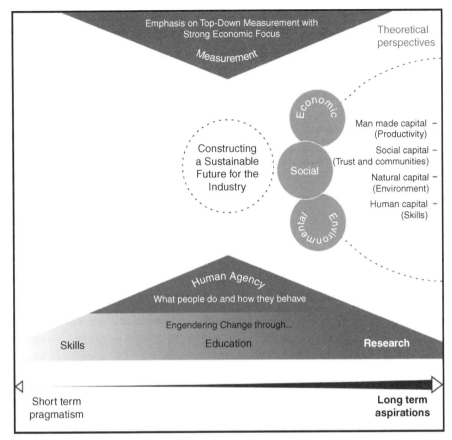

Figure 3.2 Theoretical perspectives of sustainable development in the context of the construction industry.

Moreover, the economic perspective generally views humans as rational beings, whose sole purpose of survival is to maximise economic utility. We have seen in the preceding sections how this translates only to partial understanding of such complex problems as sustaining competitiveness, skills shortages, climate change and the shaping of sustainable communities. Furthermore, such abstraction could lead to people at the grassroots feeling disenfranchised, as we argued in the case of social capital and the formation of sustainable communities. There is also the limitation of the economic perspective in dealing with the age-old problem of the free-rider. As we have seen, sustainable development often involves some level of public good, e.g. in the development of skills, the shift towards energy conservation, and the notion of the community. The economic solution may seem straightforward – a combination of free market mechanism and governmental intervention of market failures. However, the reality lies in the complexities associated with market failures and the issue of human behaviour in response to governmental intervention.

The extant literature across all four capitals suggests a need to consider, at the very heart, the importance of human agency – an aspect that is unfortunately downplayed in policy-making and much of the economics literature. Yet, the success of any sustainable development endeavour depends on appropriate interactions between socio-political and economic structures, as well as human agency,[182] and the emphasis is increasingly found at the local community level. Still, perhaps less is more in relation to the ever-expanding realm of measurement if we were to move forward in the quest for sustainable development. What is needed is policy formulation and implementation that accounts for the complexities of human agency and a deeper understanding of how these eventually shape institutional structures (and vice versa). We have largely attempted to present both the economic and sociological perspectives to the four capitals – these two perspectives often talk at each other (often to their backs). Integration of these perspectives is necessary to understand the complex interrelationships between the four capitals presented here.

The following sums up the dynamics of sustainable development neatly: 'From a system dynamics point of view, they can be considered to be independent but coupled non-linear self-organising systems. Sustainability refers to the conditions for maintaining their self-organisation processes, and investment to the measures that can be taken to strengthen the resilience and viability of the systems. Consequently, the "currency" of investment cannot be money in all cases, but has to correspond to the system characteristics. In this context, the question of weak or strong sustainability based on substitution possibilities turns out to be more of a methodological artefact of economics, resulting from single factor analysis instead of a multi-criteria approach necessary for sustainable development'.[183]

Closing thoughts

Intellectually, debates surrounding the concept of sustainable development have gained maturity ever since the term entered into policy-making discourse in the 1980s. Sophisticated models have been developed in an attempt to explain the multiple facets of sustainable development through the theoretical lens of economics, sociology and ecology. Knowledge, however, remains incomplete in terms of affording a holistic understanding of what sustainable development entails. This is in part due to the complexities associated with the dynamics of the interrelated dimensions of man-made, social, human and natural capitals, and also because the definition of such a loaded term as 'sustainability' is bound to remain forever contentious. Furthermore, theoretical developments to examine the various aspects, both singularly and their dynamic intersections, demands the mobilisation of considerable amounts of time and financial resources. Still, progress made in the theoretical understanding of sustainable development is necessary for incrementally transforming societal attitudes, behaviours and actions towards a more sustainable future.

Yet, there is still a disconnection between the aspirations of the sustainable development agenda and the reality of achieving this at the grassroots. Arguably, the lack of consensus on the definition of 'sustainability' does little to help offer guidance as to how this may be addressed in practice. Because of the existence of multiple definitions, the emphasis for policy-makers centres on refining methods of measuring sustainable development rather than its implementation. Furthermore, much measurement work focuses heavily on the quantifiable, thereby reinforcing the economic perspective that has hitherto dominated the sustainable development agenda at the expense of holistically constructing a more socially and environmentally just world. Consequently, this creates further disincentives for practitioners to properly engage with the agenda.

The lack of holism is certainly evident in our leading thinkers' responses. The chasm between academic thinking and practitioner actions on the sustainable development agenda cannot be over-emphasised. At times, the desire of academic researchers to seek the Holy Grail of resolving the conflicts between multiple facets of sustainable development can create tensions with the practitioners' survival instincts that elevate the economic imperative above human, social and environmental considerations. Pragmatically, as demonstrated by our leading thinkers, practitioners often approach the sustainable development agenda by focusing on pet passions (often framed in terms of single issues). This, however, is not to say that practitioners are not concerned with the sustainable development of the wider society. Rhetorically, our leading thinkers have alluded to the grand aim of the sustainable development agenda: that this must first and foremost consider the needs of and impacts on people in communities! However, in operationalising

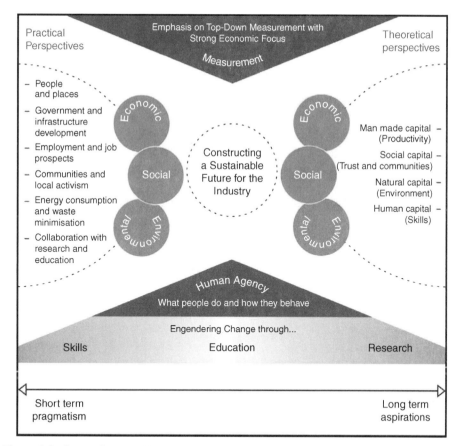

Figure 3.3 Comparing practitioner and theoretical perspectives of sustainable development of the construction industry.

this vision, they conceded that the focus of most practitioners tends to be on the immediate, on what can be pragmatically achieved in an economically viable manner as opposed to what should be done as conceptualised in the Ivory Towers of the academy. After all, what practitioners tend to grapple with are issues of the day, rather than necessarily worrying about the problems of a distant future.

Instead of concentrating on differences between the 'thinking' and 'doing' of sustainable development, it is perhaps more fruitful to consider where both might overlap (see Figure 3.3). After all, the practitioner audience is only one of many constituent parts of the construction industry, and of society at large. The future, therefore, should be more about forging closer partnerships in a number of areas. Policy-making at governmental level is one such area that is essential, as the government plays an important role as procurer, user and regulator of construction work. Here, it is important that engagement goes both ways; that policy-making remains sensitive to the needs of the plurality of actors in the industry, and that practitioners elect to act as co-producers of policies that contribute to the advancement of the sustainable development agenda. There is, of course, much that practitioners can do, and the pursuit of sustainable development need not be divorced from the perpetuation of the economic imperative. Practitioners can move towards becoming more self-critical and reflective of extant practices so as to reap savings that can then be re-invested in developing a more socially and environmentally just world. Where building future capacity is concerned, much can be gained – on both sides – by getting closer engagement with the education and academic research sectors. In an optimal world, the future might just be characterised by greater collaboration between different entities – whether this is between theoretical disciplines, or between the academy and practice, or between the social partners of the state, employers and employees – to tackle grand issues such as sustainable development. Just how this might materialise is the theme of the next chapter, when we expose the theoretical considerations and the thoughts of our leading figures regarding the notion of governance.

Chapter 4

Connecting up government, corporate and community stakeholders in governing the future of the construction industry

'In order to govern, the question is not to follow out a more or less valid theory but to build with whatever materials are at hand. The inevitable must be accepted and turned to advantage.'

Napoleon Bonaparte, 1769–1821

Chapter summary

In order to secure a sustainable future, society needs to be mobilised so that the various strands of sustainable development discussed previously can be realised. The social actors of the state, corporations and communities need to be brought together through a strong institutional framework that can engender real positive action. This chapter focuses on critical theoretical and practical issues when designing such an institutionally co-ordinated[1] response. The chapter first presents an analysis of the interviews, focusing especially on the critical issues surrounding the nature of governance structures in contemporary society. The analysis reveals our leading figures' perspectives on how the role of government in relation to the construction industry is transforming as a result of globalisation, increasing complexity of societal problems and the devolution of power to the private sector and community actors. The changes in political governance, in turn, influence the way the private sector behaves, and the chapter tracks our interviewees' thoughts on how the nature of the industry has evolved as a result of more contemporary forms of procurement such as private finance initiatives (PFI) and public–private partnerships (PPP). At the same time, the role consumers and end-users play in shaping the built environment is being accentuated, and this is all happening within the context of democratic participation of the enlightened client, the troubles with professionalism in the industry and the drive towards corporate social responsibility.

It would seem, therefore, that the sustainable development agenda has to be delivered through cooperation of the three social actors of the state, corporations and communities. The order of the day is the renewed call for joined-up thinking and working! Yet, there are inherent tensions that need to be resolved in the pursuit of interconnected ways of working. These contradictions are manifest in a number of ways. First, there is the dilemma of succumbing to the pressures of meeting short-term interests while maintaining the long-term view. Second, there is the paradox of finding localised solutions within a global context. Third, there is the tension between leading with top-down authority and coping with the vagaries of bottom-up devolution of power. Fourth, there is the question of representation and how individuals can effectively safeguard the interests of the collective. So, in developing a joined-up response to meet the challenges of the future, it would seem that grasping the nettle of these contradictions is a constant struggle.

The chapter juxtaposes our leaders' thoughts with a review of the theoretical literature on governance, initially from mainstream sources and then from the construction management context. The review traces the paradigmatic shifts in terms of political, corporate and community governance, which confirms our interviewees' observations of moving away from centralised provision of public infrastructure controlled by the governing elite (i.e. the state) towards greater empowerment of the grassroots found in the private sector and communities to deliver infrastructure development. This phenomenon is known as the depoliticisation process. However, society still needs to tackle a steep learning curve as there are many divergent ways in which collaborations between the state, corporate and community actors can be forged. Indeed, in taking an institutional theoretical approach, the nature of the institutional structures that govern positive action matters much less than the process in which stakeholders across the levels of governance align with one another. Rather than prescribe how the actors across government, the private sector and community can come together to engender change effectively, the role of researchers should be to capture the emergent lessons learnt from the dynamic configurations of the tripartite arrangement, and to examine in greater depth how the notions of power, control and authority alter over time.

The key issues discussed in this chapter are as follows:

- Interactions matter between government, private sector and community actors in encouraging positive action for the delivery of the sustainable development agenda;
- Such institutional arrangements are in constant flux, as stakeholders attempt to seek an alignment of the agendas[2] to meet the challenges of the future. As such, there is not a single method for organising the interactions between the key societal actors in forging a sustainable future for the construction industry;
- Instead of adopting a prescriptive approach in explaining institutional arrangements, researchers can profit from analysing the inherent tensions that arise as stakeholders seek an alignment of agendas. A number of paradoxes will be identified in this chapter, including the tension of satisfying short-term needs and maintaining a long-term view, the conflict between finding localised solutions in a globalised context, and the ambiguities of top-down authority and bottom-up participation;
- Institutions integrate and disintegrate over time, as a result of a rise and fall of new and old actors, respectively, involved in the production and use of the built environment. The shift in emphasis from production to consumption is reinforced in this chapter;
- There is a need for joined-up operations to be fulfilled by the intersections between disciplines and professions, and interactions developed between stakeholders across the boundaries of governance levels, departmental functions and geographic space.

Introduction

On a cold January afternoon, as the clock struck one o'clock, something out of the ordinary took place in the centre of Newcastle upon Tyne, England. Around 500 complete strangers suddenly scrambled towards the monument of Earl Grey and froze for a minute before breaking out in a conga line. This was not part of a film set. Instead, this was a display of spontaneity sparked by random users of the social networking website *myspace*. And this was certainly not the first, or even the only event known as a flash mob. Similar flash mobs happened elsewhere, including a mass public pillow fight in Seattle and disco dancing in London's Liverpool Street railway station. Indeed, the growth of social networking websites like *bebo* and *facebook* demonstrate the power of the Internet in bringing people together to do stuff, whether physical or virtual, and however trivial or significant. For instance, an 85-year-old pensioner named Peter, better known by his online pseudonym *geriatric1927*, made headline news when he posted his video on video-sharing website *youtube*, which grew out of his initial desire to combat feelings of loneliness after the death of his wife. Unexpectedly, *geriatric1927*'s videos attracted a huge fan base, particularly from youths, serendipitously bridging the intergenerational gap that has come to typify contemporary society. The Internet has also grown to become an indispensable tool in industry and commerce. Dodgson and colleagues[3] describe how Proctor and Gamble utilised the blogging concept through what they call *Connect and develop* to get scientists from all over the globe to virtually innovate and solve complex scientific problems relating to their product development.

Harvard Professor Robert Putnam, in his book *Bowling alone*, asserted that there was increasing public apathy in many of society's conventional institutions. Putnam (2000) drew from his observations of the American society and argued that its 'collapse' stemmed from declining membership in traditional civic, fraternal and political organisations.[4] Yet, the depiction of the flash mobs and the phenomenon of online social networking mentioned above imply new forms of civic participation that potentially provide a counter-argument to Putnam's thesis.[5] Indeed, Barack Obama virtually capitalised on the Internet as a new form of civic participation to garner support, especially from American youth, that led him to win the US presidential elections monumentally in 2008. Studying such phenomena is, therefore, crucially important in understanding the nature of institutions that govern contemporary society.

In the previous chapter, we highlighted the fact that knowledge about sustainable development remains incomplete. We suggested that action was needed that should go beyond political hype, and that this necessitated strong institutional frameworks. As such, we focus here on the critical issues that concern the design of institutional frameworks seeking to support positive action on achieving the sustainable development agenda. The chapter first presents the views of our interviewees on a range of issues including the force of globalisation, the changing role of government, the privatisation of infrastructure development, the engagement with end-users in shaping the built environment and the evolution of professional institutions that govern practices in the industry. One of the core messages arising from the interviews is how effective human relations matter in the delivery of construction projects, and the importance of forming collaborations between government, corporate and community stakeholders to engender positive change. In the latter half of this chapter, we compare our leaders' views against the theoretical literature on governance. The review of the literature suggests that governance can be scrutinised at three levels: political, corporate and community. In reality, stakeholders across each of the three levels interact with one another in a complex web of interrelationships, and so the chapter also calls for a better understanding of the concept of joined-up thinking.

Governance of the industry: seeking an institutionally coordinated response to meet the challenges of the future

We have established thus far the importance of human relations in any conception about sustaining a future for the construction industry. It was also argued that more needs to be done to understand human agency in the shaping and enactment of the sustainable development agenda. This necessitates an institutionally coordinated response between key actors of the state, employers and workers to mobilise meaningful engagement across the four dimensions of sustainable development discussed in Chapter 3. In this chapter, we focus principally on just how such an institutionally coordinated response might be structured by framing our analysis of the interviews. In this section, we make sense of the interviews and present a number of emerging themes that are of concern to our leading figures. Of particular emphasis is how traditional boundaries of governing and organising are constantly reconfiguring, from the dismantling of geographic boundaries in the globalisation of trade, to the increasing emphasis of private sector and community involvement in delivering public services, to the breaking down of barriers across professional, trade and disciplinary demarcations.

The section begins with our leading figures' views on globalisation. Although the advantages of globalisation in relation to broadening economic and labour markets are a welcome development, there is acknowledgement that the construction industry has traditionally been peculiarly localised, relying heavily on government investment at the national level for its sustenance. However, rapid change in terms of increased competition by foreign companies and the opportunities that emerging markets abroad afford mean that globalisation remains a significant issue for our leading thinkers. There are inherent tensions that could dampen the potential for reaping the benefits of globalisation, including the need to deal with localised concerns within a globalised context and the need for a nuanced approach to understanding diversity, both at home and abroad. Furthermore, in engaging with diversity as a result of globalisation, there is the increased burden of coordination that needs to be accounted for.

Yet, boundaries between national governments are becoming fuzzier. A corollary of this is that the role of government is constantly evolving. Here, our leading figures express specific concerns over the efficacy of governments to engender change for a sustainable future, especially given the growing phenomenon of an arm's length approach towards public governance, which sees a paradigmatic shift away from provision of public services to an enabling function. This, in turn, reconfigures the way in which construction activity is procured and demands greater involvement from the private sector and community engagement. We present our interviewees' response to the public–private interface, which is likely to remain a pertinent issue in the way the public sector commissions built environment projects on a global scale in future.

Given the criticality of the involvement of the private sector, it is crucial for us to examine how the breaking down and disordering of boundaries is associated with the managerial issues of the firm. Here, we outline our interviewees' perspectives of the management of people and relationships in construction, emphasising especially the need for cross-disciplinary working, the growing complexities of supply chains and the importance of interpersonal relationships in brokering effectively across organisations. Finally, we reflect on our leading thinkers' views on what it means to be a professional operating within the construction industry, emphasising again the need for inter-professional collaboration.

By analysing the interviewees' perspectives on how the political landscape is altering, and how business and community actors have to adapt to changing requirements all the time, we reveal inherent ambiguities of living in a world that is increasingly becoming boundary-less. The interactions in which governments, corporations and communities must engage to deliver the built environment are becoming more significant. Boundaries between stake-holders become less discrete, and it is no longer as adequate to talk about those who buy and those who sell construction products and services as binary entities. The complexities of supply chains and the complications of modern procurement approaches mean that those who pay for the built environment are not necessarily involved in shaping the design, construction, and even use, of the physical structures. Consequently, it is essential to develop new ways of understanding how stakeholders in government, corporations and communities engage with one another. We have framed our analysis in the notion of 'governance' of the industry to explore how an institutionally coordinated response to the sustainable development agenda raised in Chapter 3 might be structured. In the latter half of this chapter, we will again compare our interviewees' thoughts with developments in the theoretical literature on governance across political, corporate and community dimensions to identify overlaps and gaps for research and practice.

Think global act local

Globalisation and its impacts on the construction industry featured as a significant issue by all our leading thinkers. As Jon Rouse simply puts it, 'we are in a global economy'. Yet, it is not just economic concerns that are forcing the globalisation issue. For Chris Luebkeman, there are a number of pressing challenges – most of which have already been discussed in Chapter 3 – that demand greater attention from countries all across the world, e.g. concerns regarding the future of energy and the shift of geo-political power away from the developed Western world to emerging economies such as China: 'Ten years from now, I believe we will be over the peak of oil. Petroleum production will be past its peak or we will be in the middle of its peak. The warming of the climate will be universally recognised. The seas will have risen, and will go on rising. The high population in the Western world will continue to be ageing. China will probably be the number one economy or on its way. We will be on the way to having two super powers. So, the world in ten years is going to be a fundamentally different place'. Accordingly, globalisation stands to intensify the pressure of change exerted on the construction industry.

In this section, we present our interviewees' perspectives on globalisation. When forecasting what future lies ahead, it is inevitable that some of our interviewees maintain a sanguine outlook. Expanding economic and labour markets were identified as beneficial opportunities of globalisation. Yet, a balanced view is needed as concerns were also raised on the increasing burden of coordination in engaging with diversity that results from a more globalised world. Indeed, globalisation gives rise to a number of tensions. There is the dilemma of delivering localised solutions within a global context, which is critical as the sector is traditionally known to be a localised industry. At the same time, engaging with diversity effectively demands a shift away from 'one-size-fit-all' interventions of which policy-makers are often in favour. Thus, the section urges a more nuanced approach to fully explore the implications of diversity. Our leaders talk of the need to better comprehend what this diversity entails; whether this relates to the different types of firms operating in the industry or the cultural implications that cross-national working creates, there is a need for better understanding of the fragmented landscape of stakeholders working in the sector, both within and across countries.

Opportunities of eastward expansion: benefits of economic and social development

For many of our leading thinkers, globalisation was seen to bring about great opportunities for growing both economic and labour markets. So it follows that emerging markets, particularly eastwards, present immense scope for companies in the developed world to expand operations. At the same time, globalisation potentially encourages freer movement of people across borders, which, in turn, enables companies to tap into a wider pool of labour sources. In the context of globalisation, therefore, both economic and social dimensions of development are, again, unequivocally interwoven.

Given the significant presence of emerging economies like China, it is unsurprising that this should get mentioned during the interviews. Bob White, for instance, referred to the growth of China as an opportunity for construction companies in the developed world to focus on what they do best while at the same time tapping in to a market in need of modernisation: 'if you go to China, you've got to use some sort of the supply chain there and modernise, because just trying to go there and actually move bricks would be nonsense. [The Chinese] are very hungry for your knowledge about managing processes. Perhaps you have some good techniques that you've delivered on major projects which they have not got the experience of over the last few years. I'm a great believer in us using the consultancy strength in [the UK] because again, if you look at the balance of payments in financial terms, the only thing we export is our consultancy. We don't export any construction products at all'.

So, globalisation not only frees companies from the shackles of being restricted by geographic boundaries, but also the symbiotic relationships forged between companies from both countries mutually benefit from the industrial development that follows. For Bob, the developing economy stands to gain in modernisation efforts whereas it makes strategic sense for developed economies to focus on enhancing their core strengths. Furthermore, globalisation means that national borders are somewhat dismantled, and the free movement of labour (especially across vast parts of Europe) in turn creates a larger pool of human resources from which companies can recruit. Jon Rouse focuses exactly on the advantages of globalisation to the labour market, as he argues that 'we don't need to be constrained by our own workforce'.

For Bob White, the ability to exploit new, emerging markets seems to fit naturally into a typical economic development cycle. He is less concerned about the decline of traditional industries, as he remarked on the decline of manufacturing in a developed country like the UK: 'And people say, "Oh isn't that terrible? Let's try and do something about it, the decline in manufacturing". Why? If that's what we're good at, then stick with that and make sure that's good'. Guy Hazelhurst reinforced this point by suggesting that the UK should celebrate its capability for exporting its construction services abroad. Guy noted, 'What worries me is that we see people who are successful in coming over here, whereas we don't actually report on the success of our companies abroad. So, for example, you'll find that that there are stadia in Australia project managed and virtually built by the British'. This remark was, of course, intended to emphasise the success of British companies abroad, at a time when Australian company Multiplex ran into problems in delivering the Wembley stadium project on time and on budget.

Balancing the benefits of globalisation with the burden of increased coordination

The benefits of enlarging economic and labour markets in a globalised world are not without inherent problems. The organisation of operations across national boundaries brings with it the managerial challenge of coordination across the margins of physical space, regulatory and legislative frameworks and culture. In this subsection, we turn to the burden of increased coordination as a matter of concern for some of our leading figures.

Since the 1990s, there has been a movement towards collaborative partnerships as the modus operandi for the construction industry. It is, therefore, useful to observe that our leaders acknowledge the significance of breaking down barriers between stakeholders in what is a fragmented industry. Bob White remarked, 'And the breaking down of barriers and knowledge-sharing becomes particularly more important focus for business. And so, I think there will be a premier league of main international players who are self-sufficient'. Bob suggested that globalisation, although creating prospects for new markets and new knowledge bases, also meant that greater coordination efforts need to be expended to manage ever-increasingly complex supply chains. So, to mitigate such uncertainties and to remove the burden of such coordination, companies might revert back to employing supply chains that have been tried and tested: 'Occasionally they will use a supply chain for their own purposes, either because they're in an area where they don't automatically have access to the local industry or because they want to use their own labour etc. The reality is that sometimes before you diversify, you need to make sure that you look after the bottom line first. The problem with our industry is economic uncertainty, but at least the bigger companies have the power to deliver a portfolio of projects to weather the uncertainty. And they will stride the world and it would be great to be one of those players'. Pragmatism, therefore, accounts for this apparent counter-argument to an earlier point raised by Bob himself, which suggested that globalisation is mutually beneficial for those who venture abroad and those who receive inward investment.

For Bob, there are also coordination problems at home with smaller sized companies. He considered that the initial benefits of globalisation can only be reaped by mobilising the efforts of the larger companies, as he considered smaller firms to be less enthusiastic about the agenda. He asserted, 'I think a lot of people working in very small local environments are happy with their lot and aren't aspirational about anything to do with their sector as a whole [...] but I think we have to be honest about it and recognise what these gaps are. When you are talking about the industry doing X, be aware of which part of the industry you're talking about. It's no good saying, "We want the industry to be innovative". We don't. You know, some little jobbing builder in Carlisle or something will be thinking, "What the hell for?" He doesn't want to be innovative. He wants to put the bit of wardrobe in or whatever it is. You've got to say, "What is the purpose of being innovative?" The purpose is that part of the UK industry needs some first-class global big players. And we've got to get up there and then, when we get up and we have been recognised and got our own sort of mark and all that sort of stuff, then we'll start pushing things back down again'.

The heterogeneity of stakeholders that typify the construction industry can also be seen in the trade union movement. Alan Ritchie talked about the problems of coordinating a singular approach when attempting to negotiate a trade union response on the Wembley project, 'I'm meeting [the Australian company] Multiplex again because of the Wembley project. This is the other problem, you know, we do have other trade unions operating within the [British] construction industry, but they are general unions. [UCATT] are an industrial based union. We are a *trade* (original emphasis) union and we are subject to the industry and we will live and die by the construction industry. And, sometimes, you know, with other unions, you don't have that responsibility. It can affect statements. It can affect contracts'. So, this adds a further degree of complexity in cross-national discussions – in this case on trade union representation – because the landscape of stakeholders involved, even within a country, is not necessarily homogeneous. In a similar vein to Bob White's argument above, one needs to distinguish which segment of the industry one needs to coordinate before barriers can be broken down.

Nonetheless, Alan Ritchie recognises the need for cross-national coordination when addressing workforce representational issues, 'You see the world is getting smaller. Now, the unions have partnership with the Indians. What we have done is we have sent funds over to India to help develop the trade unions in India. When I joined the then Amalgamated Society of Woodworkers (ASW), it was us who formed the trade unions in South Africa and Australia. We financed that in Canada. We financed it in America for the joiners. So, the ASW, the union I joined, had a history of that. Now, what we have got to look at are the Third World countries, where there are construction workers being exploited [...] I do see that as an important role because, as you quite rightly say, globalisation and multinationals are coming in and I've got to try and build links with these other trade unions and see what exploitation is happening'.

Indeed, globalisation results in the reinforcement of the rhetoric of collaboration, yet increases the burden of coordination across the diverse landscape of stakeholder groups involved, both within and across countries. Nick Raynsford aptly summed this up when reiterating the point about fostering greater interactions between different types of firms operating in the sector. Nick also maintained that the future might be one that is characterised by 'Big is beautiful', as he predicted, 'I think there is going to still be a huge diversity. We may see the emergence of slightly larger construction companies in the UK. It's been paradoxical that by comparison with France and Germany, for example, our biggest construction companies have been relatively small. We haven't got many of the size of Bouygues in the UK. And I think there will be an increase in size. I think we will see, for example, in the housebuilding sector a coming together, a conglomeration of some of the separate bodies to create rather larger housebuilders. But even so, they will be very small in international comparison. So, I think diversity is going to remain a feature. And there are strengths in that, providing the objective that I've described is met, and that is that while people are trying to drive up standards in their own domain, they are not building a silo that is separating them from the rest; that they see their role as part of a wider remit. But I think we'll also see, with the continuation of the trend towards more integrated supply chains, a number of smaller companies working in regular partnership with larger ones, and as a result of that, probably both exercising a greater degree of influence but also learning a lot from the process and sharpening up their own performance'.

So, in benefitting from globalisation, it is inevitable that firms constantly reorganise. There is also a need for greater engagement across diversity, whether this is in terms of collaboration across national boundaries or in terms of partnerships forged between different types of firms. In this subsection, however, it is clear that the advantage of a globalised mode of operating requires a considered view of the challenge of coordination. In part, the increase in the burden of coordination is a result of dealing with diversity. In the next subsection, we look at the issue of diversity in greater detail as we discuss a number of inherent tensions that need to be accounted for in order to effectively cope with diversity in a globalised world.

Understanding the inherent paradoxes of engaging with diversity

Globalisation introduces the need to engage with diversity, yet in such engagement there are a number of inherent contradictions that have to be considered. Here, we discuss three fundamental issues. First, as mentioned above, there is the tension of coping with the different sizes of firms working in the sector. Second, we turn to the tension between seeking localised solutions within a globalised context. Third, we raise a question about the heterogeneity of international practices to ask whether the proverbial grass is always greener on the other side.

On the size of firms, Sir Michael Latham reiterated that different sized firms have very different concerns, which need to be understood: 'We often have to talk in such terms and the Government, of course, like to talk in such terms, but actually, it's not a single industry. First of all, it is heavily divided on grounds of size. I mean, large construction companies, like mine, employ very few directly. But the vast majority of the work, and particularly in terms of numbers of projects, as opposed to value, is in fact done by the small builders who are the ones who actually deal with the public on the whole. I mean, firms like mine don't do Mrs Jones's porches and stuff, but there are plenty of people who do and there comes the concern that small builders, who are the vast majority in terms of numbers of firms, small builders have totally different concerns'.

Unlike Bob White, however, Sir Michael Latham believed that for cultural shift to happen, improvement strategies would need to be sensitive to the peculiarities of the sector, but that all firms including the smaller sole proprietor firms would also need to be engaged. He stressed, 'For example, one of the things which I regularly have told my staff here at the Construction Industry Training Board (CITB) is, "Remember always that 98% of the 69 000 firms who are on the CITB's register employ under 50 people and 92% employ fewer than ten people". So, they are very small firms. A firm like mine, which employs about 1500 people, is most unusual'. Therefore, in engaging diversity, one needs to better comprehend the subject with which they are engaging. Although policy-makers are inclined to design 'one size fits all' policies to stimulate industrial development and influence industrial practice, Sir Michael Latham suggests that this might be futile.

Indeed, having globalised solutions may be undesirable, as the construction industry traditionally existed to serve a localised market. As George Ferguson talks of the geographic bounded-ness of the construction industry, 'I believe that architecture should respond to local character and local needs. I think the danger we have with construction and architecture is that everything has become possible and everything has become mobile e.g. in terms of moving materials anywhere across the world. I think we need to take stock of the limitations to what we can do. And I think we need to grow up to understanding that we shouldn't just use everything because it's there, e.g. being able to build a 500-metre high building just because some other country has done it. It's so not necessary and it's not sensible just because we can do it. We are not doing it because it makes a better place'. Wayne Hemingway also echoed this sentiment by suggesting that local tastes and local skills matter when designing, constructing and using the built environment, as he stressed that he '[doesn't] think you can have a central government deciding on what kind of housing is right for Gateshead or somewhere. You've got to have the skills locally'.

Tom Bloxham adds, 'If we are going to be successful in other regions, we have got to set up local bodies that can develop those relationships locally and work locally and be appointed locally. And so, we got those people and we appointed people to find somebody really talented, you know, like ourselves that had built up Urban Splash from scratch'. Guy Hazelhurst also supports the building up of local capacity, 'because I think construction, in the main anyway, is a locally delivered, locally constructed, undertaking. There are challenges and a lot of issues of culture to deal with. You have to get down to a regional and local level because that's where construction employment, construction opportunities are created'.

Guy also argued against wholesale mimicking of lessons learnt from abroad, as this often requires a nuanced level of understanding. As a case in point, Guy drew from his personal observations of the international construction market and noted two contrasting strategies for industry development: 'I think I mentioned Korea and Saudi Arabia as examples. They are just some observations there from previous lessons in the international construction development. That some countries which export business services develop much faster their own, such that, Korea for example, is now one of the strongest construction industries in the world in terms of its capacity to export building services and in a sense, it used development to lever up its own construction industry. On the other hand, Saudi Arabia, which chose to bring in the expertise rather than develop their own'.

According to Guy, these are two possible approaches that can be adopted to develop the construction industry. Given the success of Korean companies in exporting their construction services abroad, it seemed intuitive that the UK should follow a development strategy not too dissimilar from that of Korea. Yet, Guy noted that 'the same choice in Saudi Arabia is the same choices that exist for the regions in the UK'. He explained that there are similar constraints confronting both the British and Saudi Arabian industries, e.g. challenges in encouraging new entrants into the industry; hence, counter-intuitive as it may be, there are similarities in the contexts of both Britain and Saudi Arabia that, in turn, influence the adoption of development strategies. So, Guy argued that one needs to be sensitive to local peculiarities instead of importing strategy wholesale from other countries without a nuanced comprehension of the socio-economic and cultural context.

To summarise, it is unsurprising that our interviewees should acknowledge the significance of the trend of globalisation. Indeed, it is becoming a cliché to state that the world is getting smaller and becoming more interconnected. Defenders of globalisation would highlight the benefits that such a trend brings in terms of growth in trade and the expansion of both economic and labour markets. Yet, concerns were also raised by our leading figures on the increasing burden of coordination associated with a more globalised nature of operations. This can give rise to a number of tensions, including the paradox of globalisation–localisation where more emphasis needs to be placed on finding localised solutions within a globalised context. Besides, the construction industry will always have to grasp the nettle of meeting local demands as its activities and markets can be largely bounded by geography. The discussion on globalisation also touches on the need for a more nuanced understanding of diversity, and that engaging diversity necessitates a better understanding of the dynamics of heterogeneity, both at home and abroad. Without a doubt, there are potential benefits of industrial development on a global scale that can be reaped through forging collaborations between different countries across the world. The role that governments play is integral to achieving these benefits. In the next section, we reveal our interviewees' thoughts on the changing role of political governance.

The changing role of government: relinquishing control to the private sector

In Chapter 3, the significant role that governments play as clients and regulators of construction activities was discussed. For Alan Ritchie, governments have the power and duty to influence how the industry behaves, as he argued, 'What I will say is that the biggest client in the construction industry is the Government and in their procurement policy, we should be determined to set the standards [of practices] in these companies who have tendered'. Alan maintained that government legislation underscores any response that is necessary in securing a sustainable future, as he claims, 'My experience has been that unless there is government legislation to underpin that, then there is going to be a problem'.

Although it may be true that government action can shape the way industry develops over time, there is, however, no guarantee that good outcomes will necessarily be derived from government intervention. In this section, we discuss the role of government in the context of the construction industry in greater depth and explore the complex, changing dynamics that political governance is developing. We focus especially on the trend towards greater private sector involvement in the provision of public services and explore our interviewees' perspectives on the consequences resulting from this. For some of our leading thinkers, collaboration with the private sector is likely to perpetuate in the future, and, whereas this potentially steers the industry to take a longer term view, there is recognition that the rationale behind engaging private sector involvement is so that savings can be made in government budgets. As Bob White noted, governments simply 'haven't got enough money'. Care must, therefore, be taken when forging ahead with an agenda of 'privatising' public goods and strong leadership with a clear strategic direction is required to ensure that high-quality public sector infrastructure is delivered. The section therefore discusses how governments need to consider joined-up thinking and practice.

Greater involvement of the private sector in provision of public goods: a necessary evil?

The trend towards tighter fiscal power of governments, especially in times of recession, means that government budgets are constantly put under pressure. This is more so in the current global economic downturn, and it is imperative that governments across the world find new ways to innovate the public purse when it comes to stimulating infrastructure development. Sir Michael Latham explained that forming coalitions with the private sector has seemed like an enticing idea since the late 1980s, as he took us through a journey of the development of PFI: 'It's also worth remembering that PFI has only emerged over a very short time and also have done so against a background of plenty of people not wanting it. In 1984, firms like Tarmac, as they then were, now Carillion, and people like Neville Simms, then Chief Executive of it, were saying to the then Tory Government, "You haven't got any money to build roads, so if we find the money, can we build the roads for you?" And the Government then said, "No, you certainly can't", because of the so-called Ryrie Rules. Mr Ryrie was a Senior Civil Servant at the Treasury and the Ryrie Rules were that the Government can borrow cheaper than the private sector so, you know, there was no point in it and that continued to be the case for about five or six years. In either 1989 or 1990, Stephen Doyle who was then the Financial Secretary to the Treasury reluctantly agreed to some experiments in PFI in road-building. In 1992 when Lamont was the Chancellor, local public authorities were asked to be more outgoing in this to try more experiments of it. In 1994, when Ken Clarke became Chancellor, it became mandatory for them to test things of this kind against PFI. So, something which in 1984 was illegal, by 1994 it became compulsory and you know, if you are going to change the thinking of the public sector and Whitehall particularly, you are going to find some consensus along the way and that's exactly what happened'.

So, the introduction of private sector involvement in financing public sector projects seemed inevitable at a time of contractionary fiscal pressures and the need to relieve pressure on government budgets. The economic imperative that predominated during this time appeared rational as the government of the day had to seek out quick-wins to get public finances in order. However, accompanying the abdication of financial responsibility of public sector projects is the danger of relinquishing control over public assets. Sir Michael Latham expressed anxiety over this, as he explained, 'Geoffrey Rippon who was the Secretary for the Environment under Ted Heath once said to me, "Michael, you must realise that the Treasury is instinctively opposed to all public expenditure. They start from the basic proposition that all public expenditure is bad. They know that some things have to be done. You have to pay for the Police, the Armed Services and so on, but you must keep total control over it". Now, this is the problem with PFI because, of course, they have not got total control over it. They have got 25 or 30 years of responsibility for paying for something which was built last year or whatever and I have no doubt that we shall see increasing pressure coming from the Statistical Office and others who will say, "Excuse me, what are you going to do if a PFI hospital goes bankrupt? Are you going to close it down?" Because if you are not going to close it down, then it shouldn't be off the public sector account because you are, in fact, remaining as residual guarantor of it'.

In Chapter 3, we emphasised that the built environment is vital in sustaining the livelihoods of people living in communities. Therefore, there is a significant social dimension attached to any public building such as hospitals and schools. Such public assets should not be pawned solely on economic terms, as Sir Michael Latham added, 'Somebody would have to bail it out. A hospital is not like, you know, Leyland or Rover, and it would have to be preserved. It might be merged with others, but, you know, certainly, you couldn't just close the hospital down. So, I think we shall see increasing changes in that regard'. The current economic recession has certainly generated debates about investment in built environment assets, and there is no doubt that the future of private sector involvement in the provision of public services and infrastructure will come under intense scrutiny as the world crawls into economic recovery. For Sir Michael Latham, nonetheless, he sees no quelling of the tide towards more private sector collaborations in public sector construction projects. He noted, 'the trend towards the public sector being enablers and deliverers, rather than actual providers is likely to continue and I see no likelihood of changing'.

The shift in political governance away from direct provision to becoming an enabler of the delivery process is a contemporary development that is exemplified in the social housing sector. Here, we see the intensification of arm's length approach taken by government in ensuring the necessary investment in the production and maintenance of social housing. Nick Raynsford observed a 'very steep decline in the provision of social housing' and commented on the 'shift of focus away from local authorities as the main providers to a more plural framework with housing associations increasingly playing a role in providing new affordable housing'. In principle, Nick Raynsford supports such a development as he 'took a view that monopoly provision was not wise and that we needed to have a more diverse framework in terms of different providers, which is the policy that we followed in government of seeking to ensure that council housing was improved either through the council or through an arm's length management organi-sation (and ALMO), which was a concept that we introduced in 2000, or through to transfer to a registered social landlord (RSL)'.

In the preceding section, we discussed that a consequence of globalisation brings with it the development of complex supply chains that creates increasing burden of coordination. In a similar vein, a corollary of involving private sector finance sources increases the number of actors involved in making decisions about investing in public sector infrastructure. This invariably fragments the decision-making landscape further, increases the complexity of interactions and raises the need for greater coordination and joined-up working across the multitude of providers involved. Sir Michael Latham observed how the face of the industry has changed in this respect, drawing particularly from the housebuilding sector again, as he remarked, 'housebuilding has become even more divorced from construction than it was before. There are now a relatively small number of extremely large housebuilders who are building a significant part of the private housebuilding programme and I'm sure that will continue and accentuate. There has also been the large disappearance of local authority building of houses and substituted it instead with Registered Social Landlords (RSLs), with Housing Associations. My company builds a lot of houses for housing associations and we are now getting another ramification of that which is that the RSLs, particularly in the new funding regime, increasingly have to work closely not only with contractors, but also with private housebuilders. And I'm sure that that will accentuate now that private housebuilders have the chance of getting housing grants themselves from the Housing Corporation. And, I think we will also see a situation by which, just as the housebuilders have solidified into a relatively small number of big firms [...] I think we will also see a continuing and indeed, increasing flow of consolidation of the housing associations themselves'. Therefore, what began as an agenda for getting economic efficiency gains at the outset has led to a real consequence in terms of reconfiguring the nature and modus operandi of the industry. In the next subsection, we discuss the consequences of government intervention and assert that a better understanding is required of how changing political agendas can bring about unintended consequences for the industry's development.

Intentions and consequences of policy shifts towards greater private sector involvement

As mentioned above, engaging with the private sector can be a cost-effective way for governments in the short term to stimulate public sector building programmes. Referring to the sticky social problem of a shortage of housing, Wayne Hemingway suggested that the government has consciously abdicated its responsibility of provision to the private sector, as he commented, 'There's a rush to build. The government is adamant that the housebuilding industry needs to deliver more homes. There is a housing shortage'. Yet, simply framing the housing problem in quantitative terms alone is insufficient. Although assessing how many houses need to be built might subsequently drive markets and the private sector to react, Wayne Hemingway stressed that the housing agenda is broader than responding to shortages. He asserted, 'for me, the most important thing is that any home that gets delivered are sustainable, that benefits the community. You're delivering a place for the community rather than just homes'.

In this subsection, we return to the recurring theme of what success means and how this is measured, in order to illustrate how the changing dynamics of political governance to embrace more private sector involvement has transformed the performance agenda of the day. We argue that the shift away from direct provision by government means that the focus is now turned to enabling the delivery process with a concomitant emphasis on numerical performance targets. This can be seen throughout many aspects of political agenda. For instance, Jon Rouse observed the government's response to the challenge of addressing unemployment, 'I think in the 1980s, particularly following the social riots in Brixton and so on, we moved much more to an output culture. So we got to reduce unemployment. We are going to measure how much we are reducing unemployment by and we are going to reduce that by this number [... but] if frankly, the job that that person's got is actually working for £2 per hour in a sweat shop in East London, is that an output that really delivers a better quality of life?'.

Ideally, governments ought to maintain a long-term view in steering the way societies develop. But, in shifting the emphasis towards enabling the private sector to provide for public services, governments concentrate on framing numerical targets to address societal problems. Such an emphasis on narrow, quantitative assessments tends to encourage a short-term economic view, which, in turn, leads to a standardisation of solutions being proffered. Stef Stefanou also cynically remarked on the short termism that characterises government interventions, commenting that, 'it suits the government to delay building programmes because they don't want to release more money so fast. But if you mean business, you go and do it! All this business of writing reports to each other for three to four years, standardising and then forgetting about it, then somebody new comes in and changes it again'. Clearly, Stef expresses frustration over the government's failure to maintain the long-term view and lack of political will to actually deliver material change. Furthermore, pursuing political agenda framed by quantitative assessments also ignores the particularities that emerge from more qualitative understanding of the challenges confronting societies. Nonetheless, qualitative measurement might prove too time-consuming and resource-intensive for governments to undertake given the fiscal constraints.

Stef also suggested that another problem with standardised solutions is that these do not address the idiosyncrasies of the industry. For example, he referred to the UK government's attempt to regulate bogus self-employment by tightening up the Construction Industry Scheme in April 2008. He suggested that a one-size-fits-all approach to government legislation can sometimes not deliver the change intended, because 'a good-run firm does it automatically, whereas a bad firm doesn't do it'. Stef noted that 'the government keeps bringing more legislation out. Unfortunately the departments and regulators base all their new legislation on more severe restrictions, but it penalises the good firms unnecessarily, because the good firms follow it. And the bad firms ignore it anyhow. So, the enforcement of legislation is important'. For Stef, having a benign government intention is not enough; interventions such as legislation have the power to influence the behaviour of the industry, but intended outcomes sometimes do not materialise because there is a failure of governments to fully appreciate the workings of a sector made up of a diverse range of constituent parts.

Therefore, more needs to be done to examine both the intended and unintended impacts of government intervention. A historical view, such as the one promoted though the stories told by our interviewees, would be a useful start. Alan Ritchie, for instance, offered this reflection on the government's intentions to regulate the employment status of those working in the industry: 'what we have in the construction industry is we haven't been training people since the 1980s and that was when it was a previous labour government way back in the 1970s who brought in what was termed as the 714 certificate (known as the 'Lump'). And when the Thatcher government came in, they then let it mushroom because they classed them as self-employed and as entrepreneurs and it was great propaganda for them'. Alan explained that the regulation of employment status of those working in the UK construction industry had the potential to improve working conditions as it was intended to compel those who were working in the black economy to become more visible in the tax system. As time progressed, what transpired was a growth of self-employment simply because of the perceived tax benefits associated with claiming such a status, and this shift in employment status to self-employment is well known for contributing to the suboptimal performance of the UK construction industry. As the old adage suggests, the road to hell is often paved with good intentions. Surprisingly, however, there are relatively fewer studies that scrutinise the unintended consequences of government action on the shaping of construction industry practice.

In Chapter 3, we argued that one must not diminish the regulatory function of the government in developing a more socially responsible industry, as governments have the potential to exercise their power as major procurers of construction goods and services. However, the political shift towards a more short-term outlook governed by numerical targets implies that governments can sometimes collude with the industry to drive down costs and maintain the status quo ill practices. Alan Ritchie bemoaned, 'What would you want as a client? Would you want a situation where, you know, your site's working with two or three people being killed. . .bad health and safety standards? But, my experience of clients is that they don't want all of that, but they don't want to pay for it either'. Unfortunately, the government as major clients sometimes pursue projects on the lowest economic cost basis, at the expense of our social considerations. As Bob White observed, governments often 'totally undervalued the cost of building'.

The complexities of politics and the need for joined-up policy-making

We have highlighted so far a number of political shifts that have influenced the way the industry operates, including greater involvement of the private sector in financing public sector projects, the pursuit of numerical targets and cost-cutting regimes as public sector clients, and the way government legislation has attempted to regulate employment practices in the industry. We discussed how government interventions often start from a base of good intentions, often driven by the need to meet short-term demands, e.g. addressing public service provision with an ever-tightening fiscal budget. Yet, despite some of the underlying good intentions, there are often unintended consequences of government actions on the way industry responds and develops. Indeed, we have called for more research to examine the underexplored area of unintended consequences of government intervention.

Sir Michael Latham offered this perspective into how governments really operate in practice, 'I have often tried to explain to businessmen the difference between a politician's approach and a business person's approach. A problem arises, the business person's response will be, "Here's a problem. It's affecting my business. We've got to solve it. Let's solve it now and we'll do this". The politician may well say, "A problem has arisen. If I try to resolve it now, which I could do, it'll upset a lot of people and the problem may get worse. So, it's probably better if I just leave it. And then, the situation may change anyway and it may resolve itself". And that's what "Rab" Butler called, "The patience of politics"; enabling rather than forcing the issue, allowing a greater degree of consensus to emerge for the problem just to resolve itself or for changing it in a different way. The natural reaction of politicians is to try to click the issue into touch until it has resolved itself or until they can arrive at a degree of consensus for a remedy. Where things are rushed, they tend to go wrong'.

So, in building up consensus in resolving societal problems, the invisible hand of government inaction can sometimes be a plausible option. Still, the 'patience of politics' itself might just be a pragmatic choice as well. The notion of government itself has expanded beyond national borders in many cases, as can be seen through examples of the European Union and the intergovernmental efforts on dealing with grand challenges of climate change and counter-terrorism. The machinery of government as a globally complex institution itself has lent support to the fragmentation of governmental departments. To exacerbate matters, the tensions created by globalisation and the localisation agenda have also intensified interactions in policy-making at various geographic levels, including local, regional, and even European and international politics. Gaining consensus is, therefore, invariably a slow process.

As Sir Michael Latham suggested earlier, government control over the delivery of public services and infrastructure development, at least on the surface, will certainly prove more challenging in the future. In order to exert control, governments need a degree of coherence in policy formulation and implementation. Yet, gaining consensus in an increasingly fragmented world of government departments may prove challenging. As a number of our leading figures point out, there is a greater need for joined-up thinking even across government departments in the first place. In part, this need for joined-up thinking derives from the complexity of societal issues being confronted, which cannot be resolved by the workings of traditionally framed single-issue-based government departments. Jon Rouse, referring to his work at the Housing Corporation, commented on the need to instigate interactions across a range of departments in solving housing problems. These departments included the then 'Office of the Deputy Prime Minister (ODPM) primarily, but there are other government departments as well because one of the things that's amazing about the Corporation is how many government departments impact on it. You've got the Home Office in terms of anti-social behaviour. You've got the then Department of Trade and Industry (DTI) in terms of construction. You've got Department of Environment, Food and Rural Affairs (DEFRA) in terms of rural housing. You've got the Department of Work and Pensions (DWP) for housing benefit. I could go on and on actually. There are very few departments that don't have some sort of connection with housing and health obviously. You know, there are very few departments that don't touch on the Corporation's work and therefore, it's obviously pertinent that the Corporation has a policy link with each of those departments'.

Thus, the complexities of societal problems demand attention from a multitude of technical perspectives, which perhaps explains the need to foster interactions across a variety of government departments and justify the changing focus towards enabling the process of delivery as opposed to the traditional notion of government provision. However, the notion of joined-up government remains elusive for some of our leading figures. As Stef Stefanou commented, there is still a lot of confusion as to who deals with what in government, simply because there are just 'so many departments' and 'departments don't necessarily talk to each other and then change doesn't happen'. Whereas the well-documented fragmented nature of the industry can pose problems of coherently and effectively representing the views of the sector in government, it would appear that the converse is true. There is indeed scope for improving how the government interfaces with the industry generally. Nick Raynsford discussed a number of areas where joined-up thinking in government can be improved, including 'the task of improving the performance of all parts of government as a major client of construction, and that includes different governmental departments, agencies, and indeed local government'. He suggested, however, that 'unfortunately, the experience here is very patchy. There are some very good public sector clients, and we've seen some impressive procurement programmes in certain areas. We've also seen some very poor ones in other areas. So there needs to be very real improvement across the board in public sector procurement and government has a major role to play there. I think the work of David Adamson, Office of Government Commerce (OGC) Director of Smarter Construction is very important, and I hope this is going to lead to real long-term improvements'.

In order to deliver material improvements in the industry, Nick Raynsford also urged governments to take stock and clarify their strategic intent. However, such strategy formulation does not take place within a prescriptive framework. To achieve the aspirations of joined-up thinking and working in government also requires government to mature through an emergent learning process. Outlining what this process was in the UK, from a Labour government perspective, he reviewed, '[in] the 1970s it was social policy, most definitely. And I think the 1980s was about a wider political agenda, allied to social policy concerns. In the 1990s, it was very much a focus on moving from policy analysis and opposition into government and in government to try and carry forward implementation. And in the first decade of the twenty-first century, I think the overriding objective was to try and ensure we make the biggest impact out of a number of initiatives which were already there, but which need to be carried through. So, it is realising the benefits of the reform programme that have been put in place'.

So, we have noted in this section that the shift in political governance towards more private sector involvement in the provision of public infrastructure may appear to be a necessary evil in the context of dwindling fiscal budgets. Consequently, the role that governments play transforms from one of provision to that of enabling the process of delivery. In turn, it was observed that this drives government policy to be framed in numerical terms, often putting less credence on the qualitative aspects of societal problems. Such a shift encourages the government to maintain a short-term view. At the same time, societal problems are increasingly becoming more complex, and so, paradoxically, these cannot be resolved simply by government departments working in silos. There needs to be mobilisation of interdepartmental, and even intergovernmental, collaboration. Again, there are no hard and fast rules as to the right way to structure political governance, and it is critical that lessons are learnt as the emergent process of political governance unravels. It is here that more research should be undertaken to examine the intended and unintended consequences of government actions. Having discussed the implications of the changing dynamics of government in terms of political governance, the next section will explore how such transformation is affecting the private sector.

Public–private interface

Our leading thinkers have all alluded to the growth in the involvement of the private sector on the provision of public building works and services. Jon Rouse, for example, noted the sizeable contribution from the private sector on UK housebuilding as he reported, '[The Housing Corporation] are spending £1.7 billion [in 2006]. We will leave alone £1.2 billion profit for the private sector from that £1.7 billion investment. Don't say no to that because you have just created £2.9 billion of goods and services'. This, of course, is within a backdrop of a decline of public sector provision of social housing, as Sir Michael Latham noted, 'There have been tremendous changes. In 1968, for example, in that year, over 250 000 council houses were built. We'll never see anything like that again in my view. We'll certainly never see 230 000 public sector houses. This was at a time, between 1967 and 1973. I was a member of the former GLC Housing Committee and in those days the GLC itself owned 250 000 council houses. And now of course it's no longer the case, they don't do it at all. And, of course, the selling-off of council stock which began effectively in Birmingham in 1966 and continued to battle it out in the years after that and I believe it's still going on. Because it was very successful this way and certainly very popular, it moved Margaret Thatcher's Government to increasingly see public sector authorities as being enablers rather than providers and it's significant to my mind that the Labour Government has not changed from that at all. In fact, I think they have accentuated it and one will see more and more of that in my view as well'. Here, again, we see the legacy that gets continued despite a change in government, a point that was raised earlier on when we discussed government legislation on self-employment. Nonetheless, the contribution that the private sector makes in terms of plugging the quantitative gap of housing shortage should not be dismissed.

In this section, we discuss our interviewees' perspectives on the interface between public sector and private sector stakeholders to review how the perpetuation of PPP arrangements have impacted on the way the industry operates. Our leaders reiterate the importance of balancing the tensions created between the provision of social responsibility and a public good on the one hand and the short-term view driven by the profit-making motive on the other hand. Ironically, the increasing engagement of the private sector to fund, build and operate major public sector construction has started to compel firms to take a longer term strategic view. Although this may be a welcome development for industry and society, there is a constant need to ensure that the economic imperative does not supersede the moral duty of the provision of public infrastructure. After all, facilities such as schools and hospitals are not entities to be pawned in the marketplace; these bear a significant social value not to be disregarded! Collaborations between the public and private sector in the form of PFI/PPP arrangements are generally in their infancy. Both sides are likely to face a steep learning curve to ensure that policy intent matches up with the delivery mechanisms. It is here that our interviewees also talk about the need to be clear and transparent as to what the policy intent is, and how this might prescriptively shape the delivery process enacted by the private sector.

Balancing social responsibility with the profit motive

The involvement of the private sector is not unproblematic and there is still a learning curve to be had given the relative infancy of PPP arrangements. One of the most critical, if basic, challenges faced by such arrangements is the need to resolve the tensions between the ideology of social responsibility in the provision of a public good and the profit-making motive that lies at the core of any private sector business. For all our leading thinkers, it is imperative that social responsibility always precedes the profit-making motive; it is by no accident that 'public' comes before the 'private' in the term PPP! Jon Rouse argued, 'provided that it's not just rolling over to the private sector and saying, "the profit motive takes precedence", because what will happen if you do that is that people will lose trust and confidence that delivery is not enhanced. What you have to do is harness the profit motive with the social outcomes. You have to do it, harness the profit motive with the social objective, not the other way round'. Kevin McCloud also noted the power of the built environment to integrate first the 'the social' responsibility, then latterly 'the private responsibility'.

However, as Kenneth Yeang highlighted, the dominance of the profit-making motive contributed to the early shortcomings of PPP arrangements. He reflected on the failures of the PFI and explained, 'That is the biggest problem, because contractors are profit-driven, whereas architects have a higher goal, a higher vision, a higher esteem. So, that is the fault of PFI, because you know, using PFI as the process, architecture goes out of the window. Everything is seen as cost-driven'. He added, 'We do a lot of hospitals in the office. It is very difficult to work with contractors on PFI. Of course the integrity of design is being compromised. So, I think we need a new model for procuring buildings'. For Kenneth Yeang, PFI or any system would work if there is a stronger 'emphasis on design' for the benefit of human well being at large.

Referring to social housing, Sir Michael Latham also stressed that there is a need to monitor the agenda of the private sector partners to ensure that one never loses sight of the importance of maintaining public good. He explained, '[the conventional housebuilder] will still want standardised types and that's why they're likely to be looking for two or three bedroom flats if they have to build flats. That, you know, is often not what the RSLs actually need. Now, that may well be a fairly decisive factor in years to come. The Housing Corporation, under the pressure of the Government, said that money was potentially available to housing developers and they have been swamped with applications for it. Now, many of the RSLs are extremely concerned about this. Firstly, because they think that the housing developers have got more money than they have and can buy land cheaper than they can because they have more resources to do so. Secondly, they see the housing developers increasingly moving into their market and this is at the same time when there is another pincer movement going on. The Housing Corporation has made no secret of the fact that it does not want the 70 partners that it's got now, it wants far fewer than that and barely a day goes by without one reading that RSLs merging with each other and so on. And I'm sure that that will accentuate. We will see increasingly, in my view, about 15 or 20 top RSLs negotiating continually with and sometimes competing with about 15 or 20 large private housebuilders [. . .] RSLs, I think, will find themselves stabbing around for funds if the competition grows and they will have to get into bed with the house builders and that is something that to some extent has happened'.

So, if care is not taken in monitoring how the private sector interfaces with the public sector, and how the social dimension precedes the economic imperative, this will result in the dissolution of the way public goods, such as social housing, are upheld. The demarcation between public and private may become fuzzier as time progresses. Tighter fiscal budgets will be one of the key constraints that results in the inevitability of marketisation. Public sector organisations, along with their private sector partners, will constantly reconfigure their organisational structures until they align with a purely economic agenda that safeguards shareholder interest at the expense of maintaining social value.

Taking the long-term strategic view and industrial development

Nonetheless, PPP arrangements can have a positive impact on the way the sector organises its operations. Whereas greater private sector involvement has tended to reduce political governance into thinking in the short term, this has ironically enabled private sector firms to hold a longer term view. It has been known that the project-based nature of the construction industry can encourage firms to be reactive to short-term demands, as opposed to taking a strategic long-term view. However, Guy Hazelhurst observed that PFI is beginning to transform the thinking of the private sector to account for the longer term rather than just thinking on an ad hoc basis about the immediate project. He surmised, 'PFI is actually making the contractors think more about the future. There is the argument that you create proper buildings so you don't have to maintain them. So, you know, buildings that are built to last are not sustainable for business. And there is a commercial pressure on the contractors' bottom line which forces that thinking. But with PFI, I think, the industry has to think about its responsibility for the built environment 25 years down the line, rather than thinking, project by project. This will impact on culture, I hope; although we'll see what happens whether that actually gets carried over into their non-PFI projects and becomes part of the company culture [. . .] hopefully, it will take us into more sustainable development'.

Sir Michael Latham also recognised that, within the housebuilding sector, the involvement of the private sector meant that business-as-usual models are no longer adequate to cater to the rather complex and divergent market it serves, 'private housebuilders normally work on the basis of having certain house types which they build and depending on which market they are appealing to. And, you know, housebuilders are happiest building four-bedroom houses on greenfield sites in the middle of Surrey. Now that they have had to move into brownfield sites and more complicated schemes, they also need to involve the contractors as well. For example, housebuilders, on the whole, have no experience of building flats, particularly high-rise flats. . .that's not something that they do. So they are going to have to hire contractors to do it'. Therefore, the marketisation of what used to be the domain of the public sector is now making firms respond to a diverse range of needs. This can also stimulate new, innovative ways of working. According to Sir Michael Latham, there are some advantages of opening up to market forces as this might energise collaborations between segments of the industry that are not used to working closely together.

Getting public policy intent right! and the delivery will follow. . .

We previously discussed the need for clarity and transparency in policy-making. This was reinforced in the preceding section when we made the case for more joined-up thinking and working in government. In this subsection, we consider the need for clarity and transparency in the interactions between policy-makers and corporate actors given the context of growing cooperation between public sector and private sector stakeholders. Stef Stefanou certainly believed that there are benefits from his perspective in knowing precisely what they are dealing with in terms of engaging with policy-makers in government. He suggested that the agenda of politicians are not always clear, and that a lack of strategic steer from government can result in confusion in delivery.

Illustrating with an example from schools, Stef observed, 'There are so many departments in government as well who do not talk to each other. I feel very strongly about schools. If there is anything that is coming out of the government programme that can be standardised, and be made to be almost repetitive, it's schools. And yet, there are so many problems with the schools. They cannot spend what they've been allocated. And there have been three executive changes in less than four years over the school programme. And I think that's wrong. What happens now is that they cluster 5 to 10 schools and then they say the governors are responsible to the PPP consortium. So, you'll think that a good principal contractor would standardise the ten schools. But they have ten governor boards to deal with. The governors of school board are then allocated to the headmaster. So the contractor finishes having ten different clients for one cluster. And this is what makes it go kaput. And somebody has to be strong enough to say, "Look mate, this is the school, we'll repeat it. We'll obviously aesthetically change a bit, so that they don't look like the khaki uniforms, and do it". But then they are allocated to the headmasters, and of course, you have some who are good, practical headmasters, and you have some awkward headmasters'.

For Bob White, he considered the lack of political leadership to be problematic in terms of realising the aspirations of improving educational attainment. Often policy-makers in government may have started from a base of good intentions, but these can sometimes be jettisoned because of lack of thought that goes into articulating a clear delivery process. Bob explained, 'It's quite evident that this whole business of educational attainment was totally misunderstood and if we weren't careful, the whole thing would disappear into a construction programme'. He added that what the government ended up doing was offloading the delivery to the private sector, because the private sector was perceived to have 'a load of cash'. However, because the private sector is driven by the short-term profit motive, PPP arrangements are often not well thought through. As a result, some of the programmes end up with 'most of the authorities entering into the process with a shortage of cash. That will then result in them having to make do with skills they've got to repair an extension and all that sort of stuff, rather than building new schools. [...] And I think, when you get to that stage, there will be an opportunity to be much more prescriptive about what we're doing and to shove the whole issue of standardisation down people's throats much more because it's the only way to get the value for the money they have got'.

Therefore, although policy intents such as the enhancement of educational attainment might be a useful starting point, there is a need for governments to clearly think through how this can be delivered in partnership with the private sector. It appears from the excerpts from Stef and Bob that partnering with the private sector in the provision of public sector infrastructure necessitates both sides facing a learning curve in order to refine the process of delivery. Although it may be the case that the government's role has shifted from that of a provider to an enabler of the process, poor leadership could result in the public perception of government's abdication of responsibility to the private sector. The notion of control, as previously highlighted by Sir Michael Latham, is therefore pertinent, albeit in a different way. As Bob White suggested that the government as a major client should 'be more prescriptive. They've got to keep control at the centre and insist on certain things being applied to get this right'.

For Tom Bloxham, political leadership can be extremely instrumental in ensuring the success of PPP arrangements. Tom suggested, 'First of all, it will be the maverick clients who lead. You want a public sector building and the first thing that happened is that you've got a whole professional team whose main intention is to make the most money out of it. And you end up paying for scrap. All designers build things at the cheapest possible price, and after that there's no time or money left for innovation. If you spent the amount of money you spend on solicitors, put that into better hospitals, there is going to be a lot better places'. So, clients have an important role to play in terms of encouraging innovative solutions to be developed. The government, as a major client and regulator, has a significant part to play in directing the industry to move away from pursuing merely a low-cost agenda. However, it is necessary for political leadership to clearly define their policy intent, and enforce this within the delivery process.

Although the intensification of public–private collaborations are likely to continue in the future, it is also important to note that such arrangements should not be taken for granted in the future. Sir Michael Latham suggested that PPP should not be adopted for the sake of sustaining such arrangements. Instead, he urged the industry and government to reflect on procurement routes and embrace what works most efficiently and effectively. Returning to the point made earlier about engaging with localised solutions, he explained, 'I think it differs according to what the job is. I think that the PFI has gone probably as far in schools as it is likely to go. Some local education authorities, particularly, if they've got a very large programme of Building Schools for the Future, are likely to go down the route of PFI or whatever it's called, because that's what the Government have told them that they have to do. However, there is still a great deal of school building going on in certain towns with ordinary, procurement, not PFI procurement. [. . .] I was talking the other day, at a seminar, to the Director of Property of a large urban city development and which is one of the pathfinders for Building Schools for the future and he said that their instinctive feeling was that they were better off taking more traditional routes than PFI, but he had to do PFI because that's where the money is'.

So, the perpetuation of private sector involvement in financing public sector projects might well be a self-fulfilling prophecy. Before one engages in such a framework agreement, it is important for policy-makers to articulate what aspirations need to be achieved through the building programme, and determine whether private sector involvement through PFI/PPP is necessarily the most appropriate route. More needs to be done, therefore, to analyse decisions made about procurement choices. According to Sir Michael Latham, however, he believed that 'we'll see that some things will naturally be PFIs. I think that hospitals will tend to be PFIs, very large hospitals, if only because it will cost the earth to fund them through the normal system and you know, nobody's got that money. In effect, they will be paid for by hire purchase, which is what PFI is. Schools, I think, we shall see less of. Prisons, I think, are likely to be done by PFI, if they are new, but not if they are upgraded and refurbished. And in fact, the prison upgrading system which is going on now, which my own company is involved in with Lovell and some other big contractors, that is not on a PFI-basis'.

In summary, therefore, public–private partnership arrangements may seem like an inevitable development as governments, driven by contractionary fiscal pressures, adopt innovative ways of enabling investment of public infrastructure in the future. In shedding its responsibility of provision to the private sector, a perhaps unintended consequence of government action in this area has been to encourage a longer term view to be taken by construction firms. The thinking in industry has transformed from one that tactically responds to the adhocracy of piecemeal projects to one that strategically considers commercial operations over a longer term. Of course, the transition is not necessarily seamless, and the extracts from some of our interviewees suggest that both sides have to face some sort of a learning curve. From a public sector perspective, there is a need to reconfigure its role as enabler of the delivery process rather than provider in the conventional sense. This demands clear leadership and political will to enforce this in the selection of private sector partners and prescription of delivery process. From a private sector perspective, there are opportunities for engaging with innovation, whether this refers to coming up with new solutions to meet policy aspirations or to new business processes and collaborations with other stakeholders. What is certain, nonetheless, is that the forging of business relationships – in the shape of both public–private and private–private collaborations – undergoes constant renewal as they align with the agenda of the day. It is important, therefore, that one should not lose sight of the social purpose of public sector building. In the next section, we turn to our leaders' thoughts about the importance of human relationships in the delivery of construction projects.

People and managing relationships in construction

The importance of people and the working relationships forged between various stakeholder groups in the construction industry features prominently throughout this book. These issues become ever more critical given developments in the complexity of organisation in both the public and private sectors, as we have alluded to in the sections above. People matter indeed, and the construction industry relies heavily on the effective mobilisation and realisation of the interorganisational and interpersonal dynamics between those who design, construct and operate the built environment. As Alan Ritchie stressed, construction work needs to be undertaken by people and so, 'an employer is there and maybe owns the contract, but he needs employees to actually build the thing'. Bob White supported this statement by reinforcing that 'you know, at the end of the day, everything we do is about relationships. It's not about materials and structures. It's about how people can best work together and make things happen'.

In this section, we first trace our interviewees' perspectives on employment, where there is recognition of changing employment structure in the sector that influences the way in which labour is being organised. The shift towards self-employment and dependence on casualised workforce serves only to disrupt employment relations in the sector, and a plea is made for a return to the employment of localised, direct workers. The section reports on issues raised about the breaking down of barriers between various stakeholder groups, and our leaders' thoughts on the success of encouraging interprofessional collaboration are outlined here. The role that clients (and users) play in promoting harmonious working relationships across the construction supply chain is also discussed, and the section closes with our interviewees' views about the changing use of skills, knowledge and professionalism in the industry.

Shifting employment structures in the UK construction industry

Early on in this chapter, we talked about how government legislation to curb tax evasion has led to growth in the use of self-employment in the British construction industry. Furthermore, the fact that the government has moved away from provision of public infrastructure meant that the size of the directly employed workforce, certainly within the public sector, has dwindled significantly. Such a dramatic change brings about a transformation of employment relations. Traditional notions of employer control over the worker, and the trade union movement of organised labour, are thereby disrupted. The relevance of trade unions in the construction industry remains under threat as trade union density among the workforce declines year on year.

Such a trend forces new ways of thinking by trade unionists. Fundamentally, as Alan Ritchie pointed out, the role of the trade union is to ensure that the rights of workers are being protected and that 'a line of communication' is maintained between the employer and 'each individual employee'. He suggested that it was perhaps more effective to do this 'collectively through the trade unions because workers trust the trade unions as they know it's in their interests'. He added that 'the history of the trade union movement in the UK has got a great track record in defending the rights of workers, but also making workers' lives better'. However, Alan conceded that trade unions have to move with the times as well, and seek out new ways of collaborating with employers to ensure that workers' rights continue to be protected. He explained, 'One of the best examples of a good employer in construction used to be Bovis. They never liked to go into tender. Now, they go to tender and you know, the situation is that they'll find themselves bidding for the lowest price. They no longer employ direct because they don't do construction, they manage the construction process. But I got to speak at the Health and Safety Summit recently and the Director of Bovis came to me after it and said, "Look Alan, we need to have a discussion about how we can work together" and I welcomed that and the two directors came in here about two months ago and we managed to put some sort of bones together as to a framework agreement with their supply chains because Bovis no longer employs'.

Thus, because many larger construction companies no longer employ their craft workers directly, this poses immense challenges for the traditional representation and organisation of the workforce to be undertaken effectively by the trade unions. The employment relationship becomes dispersed as a result of ever more complex supply chains. Therefore, the extract above from Alan Ritchie's interview, signals the possibility of trade unions partnering with main contractors to get access to the often hard-to-reach workers further down the supply chain. Indeed, the employment structure is undergoing constant change. For example, it is unsurprising, given the dependence on migrant workers over the last decade in the UK construction industry, that new sources of labour will be recruited in the future. As Stef Stefanou suggested, 'everybody for years have been telling us we are coming to a standstill because we won't have any trades to do any work. [...] And then suddenly Europe opens up with 10 more nations, and now another three or four. And now, the trades have been supplemented by these countries, and possibly Turkey in future'. Of course, the recruitment of new sources of labour into the construction industry will bring with it issues of representation and cross-cultural working.

Nonetheless, it is comforting to note that there was acknowledgement, especially by our contractor interviewees, of the need to ensure harmonious employment relations at the construction workplace. Stef Stefanou argued, 'life is always a two-way street. You cannot expect people to respect the firm and to "kill themselves for the firm" if you don't look after them. And we will always do that. We will always be compassionate and understand things etc. We also encourage anybody – and a lot of firms do it as well – anybody in our firm who works for us and wants their daughter or son to get a job, we give them a job. Of course, some of them like it and stay; some find it's not what they wanted to do, and after few weeks or months, they go. And, at this moment, I think we have about half a dozen sons or daughters working in our head office and many more on our sites. And so we encourage this and the family touch in John Doyle [. . .] that's good because when I grew up in an environment in Egypt, and a son or daughter wanted a job, their fathers would also used to find them a job within his company'. This does demonstrate two critical issues. First, many companies in the construction industry are still characterised by a patriarchal organisational structure; and, second, the role that the institution of the family plays at the construction workplace demands greater scrutiny.

For Guy Hazelhurst, the direct employment of workers, especially from the local area, is also about building up sustainable communities. He argued that local workers, and their families, also become the users of the built environment that is produced by construction firms in the area. He questioned, 'How many firms from the North East can you get working on North East jobs? In order to do that, it has to be local firms employing local labour. How can we maximise such local employment? Because local sources provide a legacy which is an issue the industry has to consider; that local people use the built environment and one of the things that gets missed quite often is that local companies are better at creating local employment and sustain the legacy'. Bob White echoed this sentiment, as he argued, 'One client, for example, has got housing problems in the Thames Gateway. We are looking at one of their sites at the moment, whereby the developer of the housing doesn't necessarily want to employ a major contractor because a major contractor demands that he employs his own supply chain for the job [. . .] therefore, the project could be driven simply by the contractor's profit motive. What you need is management team who will employ the local community in its supply chain. By the way, while you've got it, you then teach them how to extend their activities, improve their skills into extensions, repairs, maintenance and all that sort of stuff. So, you actually create a total housing'.

Breaking down barriers through partnering in the industry: progressive or problematic?

For far too long, the industry has been criticised for its fragmented organisation. Yet it seems that traditional boundaries are constantly being broken down and that it is imperative that collaboration takes place to bring together disparate entities. Even in the trade union movement, barriers are increasingly being dissolved as Alan Ritchie pointed out how UCATT is beginning to engage with the manufacturing sector given the prominence in off-site production, 'We've already got members in the factory process side of it'. He also emphasised that relationships in construction remain in constant flux, as he asserted, 'I always see the challenge as being if the industry does change, then we will change with the industry and we will defend the rights of the workers, irrespective of where they are. I see a lot of things happening, because a lot of areas in the industry, where you as a trade union, what you have to concentrate on, what we have concentrated on in the past and what I do see is there could be a closer link between us and e.g. the designers, architects or whatever; I can see this developing because I find that the role of architects is coming more and more into say, health and safety, for instance'.

Traditional distinctions between the employer and employee are, therefore, meaningless in the context of construction, simply because of the multitude of stakeholders involved in the process of designing, constructing and operating the built environment. It is imperative, therefore, that a variety of stakeholders is engaged. Over the last two decades, the concept of collaborative working and partnering has, at least rhetorically, begun to diffuse among parties working in the design and construction process. For many of our leading thinkers, partnering is still considered to be a panacea for many of the industry's problems.

However, for partnering to work effectively, there is still more scope for forging effective relationships across the diverse groups of people working in the sector. Sir Michael Latham offered his view on a number of critical issues, 'The first strategic issue is getting a wider understanding amongst clients of what best practice procurement actually means and how to do it. The second issue, I believe, is to persuade more and more of the consultants in the profession that they are part of an integrated team and not a side subject. The third issue is to get people to understand that if they work together on an integrated basis, on partnering, as open book, then everybody will value the whole project, not just their part of it. And the fourth strategic requirement, in my view, is putting more and more effective and well-documented case studies into the public domain so that a particular client can see that schemes done this way will work best for them. And those, to my mind, are the vital issues and what will help is if there is a strong lead from Government and not just a strong lead, but a strong search light within it pressing these issues all the time'. According to Sir Michael Latham, despite exhortations of the importance of partnering, the realisation of its benefits and experience in practice remains patchy.

For some of our other interviewees, the label 'partnering' does little to explain how meaningful collaborations can be harnessed. Tom Bloxham, for example, prefers to focus on the enactment of partnering and recognition of collaboration beyond its name, 'I mean, we don't have any form of partnership, but I consider all our relationships are partnering in a way. And what I believe with any partner, you have got to have a very open, honest discussion about what every party wants out of it. We are doing a regeneration project. We want to develop the building, and we want to make some money out of it. Profit is not a dirty word'. So, partnering is not just a soft and fuzzy concept that is devoid from the hard-nosed reality of the economic imperative. For partnering to work effectively, it is crucial that all parties involved are transparent about their agendas, commercial or otherwise.

And, in fact, as Tom suggested later, there is probably more congruence across the diverse range of stakeholders than one might expect. He explained, 'Take the architects, for example. What the architects do is they want to design beautiful spaces and build their life's work…with my money. That's not a problem. That's quite good actually. But, they have got to recognise that we've got fixed budgets. So, it's not about me taking *x*% of my budget, it has to be understanding space for making profits. [. . .] Take the local authority as another example. The local authority will want something out of it. They want to bring in regeneration. They want to bring in new people to the area. They want to improve the bad points of the area and this is where my business can contribute to this as well. So, you need an open, upfront conversation on what every party wants and then we work to achieve all those things'.

Indeed, knowing the hidden agendas of those involved in partnering is the first key step in making it work. Sir Michael Latham explained, 'I was in politics for a long time, both local government and national government and parliament and I often have to talk to councillors or MPs and so on. When I do, because I was one of them myself, I never try to appeal to their best interests. I never try to say, "partnering is more ethical or more moral or more honest than", you know, "adversarial contracting", though I believe it to be all those things, but what actually, what I tend to do is I tend to appeal to their worst fears. I tend to say, "If you're the leader of Birmingham City Council, you don't want to have to get up and explain to the full Council in front of the *Birmingham Mail* and *The Sun*, why your maintenance programme is six months late and a million pounds overspent, or why the school, which you promised would be safe, is not yet completed and why it's 15% overspent"'.

Forming partnerships and collaborations is certainly not a rosy business, but if done well, can be extremely rewarding as parties involved can stand to benefit a lot through learning from one another. Bob White reflected on the Broadgate project and explained why he felt liberated in adopting, as a pioneer, construction management as a procurement route in Britain in the 1990s: 'there was no concern about hidden agendas or not sharing objectives or, you know, we don't like that, we'd rather do that and all this. We just absolutely fell in line behind everybody naturally; we didn't fight despite having strong personalities, and we enjoyed each other's company, but we did spend lots of time out of work together, but that maybe because we didn't switch off really, we never got out of work. But, so it was just terrific. On top of which, we had a great client, you know, Stuart Lipton taught us the value of buildings and architecture and good practice that we had never really been exposed to before and that was the brilliance of Stuart Lipton. He was the teacher to us there – a unique sort of vision about how best to get a difficult design built, very ably assisted by Peter Rogers, but still, working for the industry by leaving behind a legacy if you like, after all these years. So, it was good. It was a good time. Great relationships were created for a really long time, you know, that endured'.

We have discussed in this subsection the importance of collaborative working and illustrated, with our interview data, how the notion of partnerships is not a static concept, but an emergent one that changes as industry evolves over time. What is critical, nonetheless, is that neither the ideal nor the label of partnerships on their own is sufficient to break down barriers between stakeholders in the industry. For partnering to be effective, parties involved need to accept and live through a long process of working hard together to deliver projects. This demands a level of honesty and openness for parties involved in the relationship to articulate the hidden agendas that matter to each party, including how the hard-nosed reality of commercial imperative affects their operations. In return, partnerships can be rewarding as parties involved in this stand to learn from one another. Of course, the role of the client is once again considered to be instrumental in realising the benefits of fostering good relationships across the construction supply chain. Maintaining good relationships with clients certainly makes a good business case, as it helps companies secure repeat business, rather than rely on the often transactional approach found in ad hoc, one-off projects.

The need to engage with end-users

Another contemporary focus mentioned by our leading thinkers relates to the need to engage with end-users. This is nothing new really, as Sir Michael Latham suggested, that a vast segment of the industry, i.e. the jobbing builders, have always been close to the end-user: 'the vast majority of the work is in fact done by the small builders who are the ones who actually deal with the public, on the whole. I mean, you know, firms like mine don't do Mrs Jones's porches and stuff'. Bob White also echoed this sentiment to suggest that 'there's a lot of good stuff done in public consultation'.

However, several of our interviewees recognised that existing practices of consultation are not without their pitfalls. Jon Rouse suggested, 'I believe in consultation groups strongly. I just don't think we do it very well and I explained that. I think it works a lot better and our problem is that we have a tendency to consult people to death on what to consult. I actually think that strong decision-making is to consult people on how aspirations can be delivered. I think you get those two things confused, and it becomes difficult for us. You know, we talked to the people in public and they think they've got an open book set up, this creates problems afterwards [. . .] and what it means is that they're always being left out. Whereas, if you make decisions and say, "this is what we're going to do; there's the parameters and here are the choices you can make within these parameters", I think we can ultimately get a better response in the consultation process'. So, allowing users to generate their wish lists results in unnecessarily building up expectations; consequently, users feel disenfranchised when their expectations are not met because of constraints on the delivery process.

As a result, there is the perception that user consultation becomes a futile process. George Ferguson expressed concerns about this, as he argued, 'I do believe in consultation, but excessive consultation has the danger of dumbing down. I think we have to lead. [. . .] We have to show people what is possible. And, you've got to take consultation at the right time. You have to inspire. We have to be allowed to inspire, but what we've got to be careful about is that we are inspiring the right things; that we're not inspiring something that has a short life. [. . .] That's why education is a very important part of successful consultation. And so, that's why, I think we have a population that is educated to understanding and seeing what we're doing and therefore, to challenge us more to do better, as an architectural profession, built environment profession, it's in our long-term interests'. Besides, consensus can only really be reached if the parties involved in consensus-building are informed sufficiently on what they have been asked to agree to. Education of users certainly underscores the efforts by Kevin McCloud to 'get people talking about the built environment, to claim public spaces, and to claim ownership of the built environment'.

The provision of choices is also what Wayne Hemingway stressed when he discussed what effective user consultation meant. Far too often, he contended that 'you get more stakeholder involvement if there's choice; at the moment, there's not a lot of choice. When there's not a lot of choice and when there's a shortage of housing, then the public doesn't get choice. So, you know, there has to be more competition and that creates more of a choice. At the moment, you know, if you want to go out and buy a car, you've got a massive choice and therefore, you push the car manufacturer further and further to come up with a better and better and cheaper product. But, the housebuilding industry isn't like that, so there's not really a lot of stakeholder involvement because they don't need it'. Just as George Ferguson argued, engaging with end-users effectively provides the impetus for the industry to keep developing innovative solutions, and, in the long run, this will be beneficial for industrial development.

Bob White, however, questioned whether we really need to engage the public on everything. He argued, 'I totally believe in public consultation, in terms of things like master plan and what you're going to do, stock transfer of council houses and all this sort of stuff, but I am a bit wary of this thing about the industry not being close enough to clients and close enough to consumers and close enough to the public, in the sense that most successful industries aren't that. GlaxoSmithKline, who we all use because we take an aspirin or whatever you take every day, don't get close to me. I wouldn't be upset if they don't get close to me as their client. [...] I think there is something wrong about this obsession that some people have suddenly got about the problem with our industry is about our reputation, our status and how people perceive us to be. It's not important really if I'm a bastard, as long as I produce the product and people are happy, you know'.

What Bob White is concerned with, in a similar way to the concerns expressed by George Ferguson, is that professionals allow the uninformed public to determine the services provided by the industry. Yet Bob White asserts that it is important for professionals to know what they can offer in terms of the products and services that they sell, and be good at delivering these. As he noted, 'there is a balance to be had there. I think, at the moment, because I'm just so good with some of the users in defining what it should be doing and where I should be doing it and all this sort of stuff, it grabs on any little thing. It's like if you say, "'Oh, yes we're now doing this and therefore, we will be better'". For Bob, jumping on the contemporary bandwagon will not move the agenda forward; whereas being a conscientious professional who delivers a product that end-users are happy with will. It is here that mobilising the skills and knowledge of those working in the industry is key to effective development.

Mobilising skills and knowledge in the construction industry

So, managing relationships in construction is complex given the myriad of interactions that exists across the supply chain, government, client and end-users. To manage the dynamics of these interpersonal and interorganisational relationships, demands a new skills set that transcends traditional disciplinary boundaries. The need for interdisciplinarity in construction was certainly a point raised by a number of our leading thinkers. At the crafts level, Guy Hazelhurst talked about the role of multi-skilling, 'I think there's certainly more scope for multi-skilling and we are moving more from what I call transformation to assembly. So, multi-skilling is found essentially everywhere, particularly, in repair and maintenance, where you get a tradesperson who can do electrical installations and plumbing works etc. One of the arguments is that a well-organised team of specialists, who can use the skills productively and this is where project management comes in, and sometimes, better management of single-skilled labour can be just as, if not more, effective'.

Indeed, craft skills in British construction have been known to be fossilised in feudal times, and there is now a progressive agenda for developing skills for modern methods of construction. Guy explained, 'There is this issue as well which is around modern methods of construction, which is more about the skills that you need in terms of making sure that the interfaces work. I just think that construction is actually about interfaces. That's what construction is. It's not just the difference between one brick and another brick, what you do between mortar and a panel or another panel or a course and another course. So, it's where the interface is going to come together between trade and trade or material and material. The things that are, as I say, about transformation and assembly, and building more off-site and balancing the traditional skills on-site in a precise and engineered way'.

Indeed, multi-skilling is increasingly becoming the norm at the workplace. Sandi Rhys Jones, for instance, commented in her interview, 'I discovered a few months ago that the receptionist and office manager here in the London office trained as a structural technician in her native Lithuania. It seemed logical to make more use of her skills, so now Simons Design in Lincoln sends her drawings and assignments over the intranet and she is now doing what she is trained professionally to do as well as managing this office'. Of course, the claim made on multi-skilling might just be a disguise used to hide the real agenda of exploitation. Yet, in a sense, multi-skilling can be appealing to the individual as well, as it enables them to develop skills to their full potential. And Sandi certainly felt this is where traditional disciplinary boundaries can be somewhat prohibitive.

So where multi-skilling appears to be a welcome development for the industry, our leading thinkers also commented on the need for more interdisciplinary and interprofessional working. Chris Luebkeman noted the importance of melding the sciences and the arts, as he reflected on his own education, firstly in engineering and latterly in architecture, 'At Cornell University, the campus is designed so that the architects are on one side of the campus and the engineers are on the other side of the campus. So, in order for the architects to attend the class in the Engineering School, you have to literally walk all the way across the campus and vice versa for engineers who attended classes in the School of Architecture. So, it's a big conscious trek'.

But it was Chris' postgraduate education when he had his first appreciation, as an engineer, of the building form from an architectural perspective. He recounted, 'I decided to return to the university, the ETH to do a Doctorate in Architecture. During this time when I was going back to study architecture, I had become very good friends with two individuals, a guy named David Bushnell and Daniel Huiden who were both graduates of Cornell University and they were doing a professional year out, if you will, over in Switzerland. They were very interesting because they also taught me how to sketch. Because at the very beginning, I had learned engineering drawing, engineering drafting, which is different to drawing, very precise and very exact and you have an ink rapidiograph with 0.25 mm. As I say, my early sketch books were just these terrible shaped corners because when I went around looking at these buildings with Daniel and Dave, I was looking at technical aspects. I was looking at the materials and trying to figure out how something has held up, and I was looking at the structures and we were having very interesting discussions about domes and vaults and the way in which you would have different layers of face and layers of structure and they were explaining to me, trying to help me understand this layering of space. And I would be trying to help them understand the layers of structure'.

This interdisciplinary thinking also featured in George Ferguson's interview, as he suggested that architectural education requires both 'literal literacy' and 'visual literacy'. However, he complained that the way we constitute professionalism in our sector tends to compartmentalise what professions should or should not be doing, usually in narrowly boundarised and fossilised terms. Sandi Rhys Jones also commented on the battle between the sciences and the arts, as she observed, 'There was a very clear definition between arts and sciences and in fact, when it came to educating my sons, all three showed the same mix of arts and science interest, an ability that clearly poses a problem in some schools for some children'. Her solution was to look 'very seriously at the International Baccalaureate, for one of my sons because I wanted to find some way of harnessing both sides of the brain. And I was also at school at a time when there were a very large number of us. I was a typical baby boomer and I became very frustrated and my efforts to get into some form of technical or scientific area were at one point actually, specifically blocked'. So, even in deploying the skills and knowledge necessary for a broader understanding and appreciation of construction, we see the dismantling of boundaries, whether these are boundaries relating to feudally defined craft skills or the boundaries between disciplinary knowledge. In the next subsection, we turn to review our interviewees' thoughts about what it means to be a professional working in the construction industry.

Professionalism in the construction industry: what's in a name?

There is no doubt that continuous improvement featured quite highly in the minds of our leading thinkers. Every interviewee considered that the future of the sector will be one of increasing professionalisation. However, a strong distinction was made between the need to be good in one's profession (job) and the organised form of professionalism characterised by professional institutions. Stef Stefanou explained that you needed both to co-exist, as he explained, 'a good company, be it a specialist, be it a principal, be it whatever, has to have a mixture of middle management and upper management which comes from two avenues: one avenue is graduates, what you call "Professional people"; the other is what you call trades. And there have been some very good examples of people who have reached the top and have been very successful, and who understand their company. And I think successful companies, or innovative companies should have a mixture, because the graduates can learn from the trades' guys and the trades' guys will learn from the graduates and it will make a very nice mixture'. For Sandi Rhys Jones, being professional, 'with a small "p"', simply means being good at what one does; she put it simply, 'I don't want an unprofessional plumber under my sink at midnight. I want a professional plumber'. Nick Raynsford also noted that to become a professional industry, the vast majority of the sector, i.e. the small jobbing builders, would need to raise their game in terms of offering high-quality goods and services to the clients and users they serve: 'I think the smaller companies will need to boost their image as professional, as high quality, and value for money, and that schemes like Trust Mark, I think, can help in that respect'.

For Stef Stefanou, becoming more professional seems inevitable if businesses want to differentiate themselves from the competition. He observed that with the perpetuation of subcontracting and the increasing complexities of supply chains, the industry needs to seek professionalisation at the bottom of the chain to ensure that those who actually undertake construction work become experts at what they do. He reflected on his own experience, 'In the 1960s, we were called the "subbie". In the 1970s, we were called the subcontractor. They gave us a bit more status. In the 1980s, we were called the package contractor. In the late 1990s, we were called the specialist contractor'. Notwithstanding the label – and the label appeared to confer some sort of status – this represents, for Stef, a significant shift in the industry over the last 40 years, and further justifies the need for professionalising the industry as a whole. He explained, 'when I first joined [in 1972], very rarely was there subcontracted work, and now it's the fashion. Now, I have a theory at that time, that the reason the main contractors were subcontracting, they were just passing on risk and the conditions of tax etc. to other companies, not having to worry with grey hair. And I immediately realised that the future of the subcontractor is to become more professional. Now most subcontractors never had engineers running the works those days, even the groundworks, the substructure etc. they used to have a foreman who was a very good construction man and they used to represent [the engineers] on site. But they didn't know much about programming, about resourcing, about talking to clients'.

So, for Stef Stefanou, becoming more professional means retaining one's core skill, but supplementing this with skills that enable one to work effectively across the design and construction process. Guy Hazelhurst also reinforced this by suggesting, 'I think we seriously have to professionalise and we have to make sure that there's no complacency. That there is a risk we are not actually losing that modularity in education, but we may be losing in the core skills in terms of engineering, of inquiry and investigation'. Guy added that this is where professional institutions have an important role to play when they accredit educational courses, 'I think the other thing that needs to be addressed is the link between the institutions that accredit higher education courses and the employers who are the end-users of the product. And I think a better link is required. There is a need to ensure that the quality of those coming out into the industry is actually useful. So, I think there is an issue about, you know, in terms of some of the people who are buying it. I think the professional institution accreditation officers that are involved are not used to working in all areas now, but there is a need for a wider education. So, I think this is something that needs to be addressed in the school system'.

In Britain, however, being professional as defined along the lines of professional institutions implies a boundarised way of operating, which still subscribes to the silo mentality. Sandi Rhys Jones maintained that professional institutions are very good at keeping people out. For Chris Luebkeman, nonetheless, maintaining boundaries may be a necessary evil, as he commented, 'Well, you know, Robert Frost said, "Americans who have good fences, make good neighbours"'. Still, Sandi argued that there are benefits of bringing down barriers between the different groups of people working in the sector, as she observed, 'it was very interesting just watching the people and the different languages in construction. The client speaks a certain language. The contractor has a language. The professional, who is the surveyor, the architect has another language. I mean, "a contractor has a customer; a professional has a client". They have specialist subcontractors and they all speak a slightly different language'.

Hence, the idea of professional institutions is like the Tower of Babel, and the fragmented, often adversarial nature of construction gets reinforced by professional boundaries. Nonetheless, Chris Blythe recognised this crisis with professional institutions, 'I think professional institutions have lost their way, frankly. And I'm talking about them all, not just in construction, but generally. And the implications of this, they have gradually lost their power. They have lost the public respect. If you go back to the origins of professional bodies, they were set up in the nineteenth century when the British Empire was developing in the industrial revolution. I think they were formed by groups of people who were given the privilege, but who were doing a few things in the public interest. [. . .] They were given privileges, but the chartered status also gave them a whole load of responsibilities. And I think it worked quite well at the start, but I think what's happened is that the institutions as they evolved today have taken the privilege and protect them, it's elitist. Some people think that the institutions are for members' interest, and not for the public good. And I think that's where it's lost the plot and if you can trace it back, they have just not been able to adapt after the war, with the rising tide of globalisation of business, the speed of communication and all of that. So, what's happening is the profession has the right to set auditing standards and all that bureaucracy'.

Jon Rouse agreed with Chris Blythe's assessment, and added, 'Professionals in this country are undervalued. I think we underappreciated technical skills. We're always beating people up, we're always beating architects up, always beating construction people up and we underestimate the difficulties of the different projects and we are too ready to criticise and I think we have got to be a little more trusting. For professional institutions, there is an issue of accountability and accountability is very important and that's different from an alliance'. Despite Jon's sympathy for professionals, he also recognised that 'we have stopped the pendulum. We have swung too far the other way to the point where we dumb down the professionals and undermine it that in the end we have lost the credibility. In the construction context, the major backlash that was still ongoing from the 1960s systems buildings, and the fact that that was blamed as unprofessional, was a myth. It wasn't an entirely professional failure; it was actually a political failure, in terms of financing it and following through'.

Nonetheless, public perception matters in terms of how professionals are regarded in society. Guy Hazelhurst suggested that professional institutions are often perceived as being, '"an old boys" club. I think one of the big problems in construction is the professional institutions in the main create barriers; they are there to earn their subscriptions'. Indeed George Ferguson noted that it is in the professional institution's 'long-term interests to have a population that's educated in what we do' so that membership dues can be paid. Bob White called this 'the strongest professional conspiracy in the UK than anywhere else in the world'. George also suggested that professionals sometimes tend to deviate away from the main focus of the service they should provide for the wider society at large. Speaking about the architecture profession, 'I think there has been a fault in our profession that architects designing for architects too much. Yes. Too much I think, and that has led to a star system (see Box 4.1) of some very spectacular architects who were doing some very spectacular architecture which then influences the schools because the students all want to be like these architects. Well, what we should be thinking is the majority support of the people and the places'.

Box 4.1.   Kenneth Yeang's typology of what a professional architect means

Delving on the architecture profession, Kenneth Yeang proffered a typology of the main types of professional architect. Although these relate specifically to the architecture profession, these are equally applicable to other professional groups working in the sector. The following types of architects demonstrate the diversity of knowledge bases mobilised in a professional working in the construction industry:

- Architect as a star designer, as a signature designer;
- Architect as a super expert, so you can be a Jon Jerde-type architect and become an expert in shopping malls design for instance;
- Architect as a project manager;
- Architect as a producer, who brings different people together;
- Architect as the developer;
- Architect as a businessman and entrepreneur;
- Architect as a teacher; and
- Architect as a service provider.

Kenneth maintained that 'the problem is that presently 99.9% of architects are service providers. And you have to move up the food chain. The idea with being a service provider is that you constantly push for fees becoming lower; you become more competitive, you get architects who become more aggressive and compete with each other'. And he argued that this is neither aspirational nor sustainable.

So what role do professional institutions play in contemporary society? Bob White asserted, 'I think that what we should do is integrate, work together and I accept also there's a real need to have professional contractors to do specialist work and you have to be extremely professional'. George Ferguson also touched on the recurring theme of interprofessional working, and suggested, 'I think that architecture has got to engage much more with all our fellow disciplines who are involved in the economy and I think we should do that much more and I would see education taking that lead. So, you don't just go to architectural school. You go to the faculty of the built environment, where there's much more cross-fertilisation of disciplines'.

Bob White also indicated that interprofessional working might just help improve the way schemes such as PFI are being approached in Britain. He noted the historical dominance of the professions in driving construction projects in the UK, and suggested that cross-fertilisation between the professions and the contractors might not be a bad thing, 'The interesting thing about the UK is: it is one of the most innovative parts of the world industry. I mean, you probably get more one-off buildings in the UK than any other part of the world and why is that? It's because it is the professions that drive projects in the UK, not contractors'.

Nick Raynsford also supports the need for professionals to share knowledge and practices to improve the industry as a whole. He remarked, 'Absolutely, and the Construction Industry Council (CIC) is a multi-disciplinary organisation that brings people together from all the professions. And we believe very strongly in promoting a multi-disciplinary approach. Now, that needs to go even further, because the CIC is one of the umbrella bodies representing the professionals in the industry, but there needs to be a similar sense of working together with the construction confederation who represents the major contractors, the clients' forum particularly the public sector construction clients' forum which is very important to pull together the construction clients, and the materials producers. . .all those involved in producing equipment and products'.

Chris Blythe commented on the future of professional institutions, 'What does it mean now? You know, when you watch the football on the television on a Saturday night, you hear about the "professional foul". The word "professional" is used loosely. [. . .] The other thing is I've come across a lot of people in this industry who call themselves professional, but they don't actually pay any fees or swear to the rights and responsibilities of professional institutions. I think, "professional" is about people who do what they do really, really well and there needs to be a distinction. [. . .] I've known some members working in construction industry who are members of the CIOB because of the salary. But some of those who are not members of the profession, you know, they are conscientious. They do a very good job. They work to a high standard and probably better than some of the guys who have professional membership. Well, I think we need the professional bodies to go back to what they were doing in their charter and start thinking a little bit more about the public interest and maybe less about members. [We] must be seen to serve public interest'. He stressed that professional institutions are not about maintaining 'the old tradition of the elitist. And the world has moved on. [. . .] It's all very well, saying, "Oh, that is our tradition". But, if the institution is so traditional that it doesn't change for the future, it will disappear. After all that history, it will be gone'.

Bringing interactions to the fore: exploring the intersections between government, corporate and community actors

In making sense of our interviews, the fundamental message that is reiterated by our leading figures is that interactions between people matter in the production and use of the built environment. Yet our understanding of how human interactions can be organised becomes disrupted as traditional boundaries continuously get disordered. We have seen this expressed in our interviewees' thoughts on globalisation, the melding of public sector and private sector interests, the breaking down of barriers between disciplinary and professional silos, and increasing engagement with the consumers of the products and services generated by the 'industry'. It is more than figurative to state that the world is getting smaller; the growth in mobility of capital and labour across boundaries, geographic or otherwise, necessitates a better understanding of how collaborations can be accomplished effectively.

Again, there can be nothing prescriptive derived from such understanding, as configurations of human relations remain in constant flux over time with the rise and fall of new and old actors, respectively, involved in the shaping of the built environment. For example, we have seen how greater emphasis has been placed on increasingly complex supply chains, with the perpetuation of subcontracting. It is no longer relevant to simply scrutinise the operations of the major contractor as projects are now delivered by a multitude of companies, often straddling across industry boundaries (e.g. manufacturing and services sectors). Rather, it is more critical to shine the light on the interorganisational and interpersonal dynamics that matter in these interactions. End-users have also been acknowledged by our interviewees in the role they play in terms of defining the nature of the built environment that will be developed in the future. Although our interviewees recognised the importance of consulting end-users, there are divisions as to how the consultation process should be directed. On the one hand, some consider that end-users should be transformed into informed participants in shaping the process and outcomes of the design and construction of the built environment. On the other hand, defenders of the professions indicate that those working in the industry should know better than let users dictate how the end result should be achieved.

In any case, the recognition of the significant role that supply chains and end-users play in the delivery of the built environment does not mean that these actors have never existed before. Instead, the lack of emphasis thus far simply implies that their role has hitherto been downplayed in the industry's discourse. Their emergence, therefore, signifies a changing trend that is likely to accentuate in the future. This trend is twofold. On the production side, the growing complexities of accounting for finance and risk in designing and constructing the built environment has led to the development of complex supply chains mobilised in the sector. On the other hand, the promise of user engagement corresponds with the rise of consumerism in the 1980s, such that end-users are becoming more empowered to lend their voice to the production process. Of course, the challenge remains as to how these new interactions can be better harnessed. Researchers have often diligently produced models and tools to illustrate how collaborations can be managed. Yet, as the landscape of actors in the industry alters, and as roles undergo constant transformation, it is perhaps more useful to examine the learning that emerges through the interactions, as opposed to prescribing partial, rose-tinted solutions that often do not mirror reality. In so doing, focus could be placed on examining the implications, both intended and unintended, of the interactions between stakeholders involved in the delivery of the built environment.

A number of contradictions have been observed in our analysis of the interviews. First, the relinquishing of public sector control and the growth in private sector involvement in financing construction projects has given rise to both opportunities and threats. Paradoxically, the collaboration between public sector clients and private sector firms has constrained political leadership to think about the short term on the one hand, while at the same time, enabled private sector firms to start thinking more strategically about the long term. Of course, how the latter has filtered down through the small and medium-sized enterprises (SMEs) operating in the supply chains remains a relatively unexplored area.

Second, there is contradiction associated with the dynamics of globalisation. Whereas globalisation has encouraged relative freedom of capital and labour mobility, the emergent tension of balancing this with a sense of localism in the production and use of the built environment meant that our leaders considered the importance of thinking about the global context, while emphasising the need for localised solutions. Third, globalisation probably also accounts for the complexity of government intervention. Ironically, although our interviewees alluded to the trend where the nature of arm's length political governance implies that governments are becoming more inclined to frame societal problems and policy interventions in simple numerical terms, the grand challenges of ensuring sustainable development demand the need for greater interdepartmental and intergovernmental cooperation. So, whereas there is a disintegration of the role of government in the provision of public infrastructure on the one hand, there is also a pressing need for integrating the functions of government, framed in the rhetoric of joined-up thinking on the other.

Fundamentally, the social dimension appears to feature prominently in our interviews, with our leading figures placing much credence on discussing how the industry at large needs to consider what is good for society. So, whether it is about explaining how good business public–private partnerships are about delivering timely and effective services to the public, or whether it is about defining the role of professionals in contemporary society, our leaders acknowledge the importance of social responsibility in whatever the industry does. This consequently influences how one perceives the shaping of the 'industry'. The changing dynamics of interactions between stakeholders operating within the 'industry' means that one needs to take a broader view that considers the impacts of cooperation between government, corporate and community actors. In Chapter 3, we discussed the need for an institutionally coordinated response to meet future challenges of sustainable development. Just what this response should look like is likely to remain elusive. Instead, research efforts should be re-focused to describe how the endeavour towards forging more effective interactions between these three societal actors evolves as time progresses, and how they respond to the inherent tensions created by societal change over time. In the latter half of this chapter, we examine in greater detail the theoretical literatures relating to governance by delving into scholarly thinking about the development of interactions between governments, corporations and communities to see how this is applicable to conceptualising the governance of the construction industry.

Shifting perspectives of governance

The term governance has traditionally been associated with public administration and the ability of government to steer society.[6] However, this conception has seen, in recent times, a broader shift towards including (and emphasising) the role of such other societal actors as the private sector and community groups to solve problems and influence outcomes in society. In this section, we critically appraise the shifting perspectives of governance to examine how government, the private sector and community groups interrelate in the networked world that we live in today. We first review the theoretical literature on governance, exploring this concept across three levels of political governance, corporate governance and community governance. In so doing, we reveal how more emphasis is being placed on corporate and community governance as governments are perceived to move away from a role of direct provider of public services to one that merely enables the process of delivery. Following on from this, we chart how some of these ideas have cascaded down into the construction management literature.

Political governance: governance without government

On the face of it, governments appear increasingly ineffective. The ability of governments to wield control over a range of contemporary issues such as the preservation of security, the avoidance of the credit crunch that led to the global meltdown in the financial system, the management of immigration and the tackling of climate change seem to come under increasing pressure, let alone their ability to steer society as visionaries. Talbot, in reviewing the efficacy of British government agencies, suggests that government 'departments have also had problems developing strategic management of themselves [...] so it is hardly surprising they find it difficult to apply these disciplines in their relations with agencies'[7] that execute the operational functions of government. The public perception of government incompetence is further reinforced by problems (and at times, failure) of government projects. Well-documented delays of high-profile building projects like the Sydney Opera House and the Scottish Parliament building in Edinburgh continue to be presented as case studies of how not to undertake the process of government-led projects.

Furthermore, if we were to take the global hype of sustainable development that we discussed previously, despite years of legislation and regulatory attempts, there is still evidence that cultural change at the business and individual levels remains at best a holy grail. The effectiveness of the Landfill Tax introduced by the UK government, for instance, to alleviate levels of wastage by businesses remains questionable.[8] More recently, the UK climate change levy on businesses has been seen by industry as another government revenue stream without bringing any real change to behaviour. Hansford and colleagues, when reviewing industry's views on the climate change levy, concluded that businesses were often left feeling 'frustrated at the lack of commitment at the highest level within their organization' and added that 'In order to achieve the government's targets the impact of the levy and commitment to a reduction in the consumption of energy need to be part of the responsibilities of senior members of staff'. They found that 'At present [the climate change levy] is seen as another bundle of red tape to be dealt with by junior and middle managers', and suggested that 'This behaviour may be explained by the piecemeal global approach to the requirements of the Kyoto agreement'.[9] This observation certainly resonates with Stef Stefanou's perception reported in Chapter 3 of a lack of political will and joined-up thinking and working in addressing the sustainability agenda.

Peters and Pierre coined the phrase *governance without government*[10] to signify limited capacity of public administration in dealing with the changing dynamics of societal needs. They asserted, 'The idea that national governments are the major actors in public policy and that they are able to influence the economy and society through their actions now appears to be in doubt. Some of the strain on national governments has been the result of the increased importance of the international environment and of an arguably diminished capacity of those governments to insulate their economies and societies from the global pressures'.[11] Nonetheless, they stressed that shifting the role of governance away from government was a natural step in societal development, given the networked world that we live in today, ranging from the enlarged European Union to the proliferation of non-governmental organisations (NGOs) to blogging and online social networking. Indeed, Peters and Pierre surmised, 'Perhaps the dominant feature of the governance model is the argument that networks have come to dominate public policy. The assertion is that these amorphous collections of actors—not formal policy-making institutions in government—control policy'.[12] We see this development in many facets of public policy, from the intergovernmental cooperation and the power of NGOs in influencing actions to mitigate the effects of climate change, to the use of business networks to supplement public provision of vocational skills training and development (see Box 4.1[13]), it would seem that implementing many areas of public policy cannot do without the cooperation of other societal actors found in the private sector and communities.

The integration of other stakeholders outside government to deliver public policy does not, however, mean that governments have lost their power. Rather, the nature of government authority and the control that governments can weld in the provision of public services are changing. For Geoff Mulgan, political advisor to former British Prime Minister Tony Blair and proponent of joined-up thinking in government,[14] this is an inevitable development given the trend of globalisation and the resultant need to involve intergovernmental coordination of public policy; furthermore, there is also the trend of democratising public policy downwards to people in the communities, the private sector and media. Mulgan, however, maintained, 'that the myth of powerlessness is one of the optical illusions of our times [...] the basic capacities of governments have not diminished. The capacity to tax, for example, remains in rude health. Across the OECD governments' share in GDP has risen not fallen over the past few decades; even the tax take (as opposed to the rates) on profits has risen. Competitiveness rankings show that many of the world's most competitive economies are overseen by relatively big governments (how they spend matters much more than what they spend)'.[15]

As Peters and Pierre argued, the role of governments has shifted away from direct control to exerting the capacity to influence.[16] Rather than being judged as weak in terms of control of society and its resources, it is therefore critical that the role of government adapts to meet the ever-increasing phenomenon of self-organisation in society. Caporaso and Wittenbrinck suggest simply that new modes of governance '[...] are based on procedures that are voluntary, open, consensual, deliberative, and informative'.[17] The transformation of political governance, certainly in the developed world, means that the nature of authority exerted by political leaders is also being reconfigured. According to Caporaso and Wittenbrinck,[18] the forms of authority can be classified in the following categories, and these suggest greater engagement with private sector and community actors:

- State-centred, political form of authority: government conventionally 'dictates' in a top-down fashion;

- Expert authority: decision-making by expert committee;
- Private, market-based authority: use of privatisation and market instruments; and
- Popular authority: emphasising the importance of referenda and the relevance of public information and consultation.

Arguably, 'government knows best' is an outmoded concept, and there is greater scope for wider consultation with the private sector and the citizens in general, as well as engendering learning in policy-making whether this is between government and their electorate or across countries.[19] The government's role therefore shifts from one of state-centred authority to that of expert, market-based or even popular authority. Such shifts in political governance make it easier for the state to blend public and private resources to achieve good for society. Indeed, 'It appears that whatever the State does it does poorly, while the private sector (for profit and not for profit) is more effective'.[20] Governments across the world are typically large bureaucratic machines. Such colossal organisations can be clumsily inefficient.[21] By contrast, the private sector has traditionally been perceived to provider a leaner, more efficient approach to resource utilisation and problem-solving. The global financial crisis that has recently crippled many nations across the world would certainly test this assumption, and calls have been renewed for a more reflective debate on the nature of corporate governance. At this point, we examine the role of corporate governance in view of the emergence of corporate social responsibility.

Corporate governance: the rise of corporate social responsibility

Just as political governance is about the nature of power and control between the state and the society it governs, the field of corporate governance examines power and control in relation to corporate organisations. At its heart, corporate governance is concerned with issues of accountability and control over the firm.[22] Shleifer and Vishny, in a seminal review of corporate governance literature, began by stating, 'Corporate governance deals with the ways in which suppliers of finance to corporations assure themselves of getting a return on their investment'.[23] They explained that corporate governance is simply about the mobilisation of financial capital, and its aim is to ensure that those who have the financial capital to invest actually get their money (and profits) back from the businesses in which they invest. Therefore, it is largely concerned with controlling managerial behaviour, through a complex regulatory web of economic, legal and political institutions, to prevent managers of firms from making bad decisions on investments and resource allocation, as well as to avoid such inappropriate actions as corruption and embezzlement of funds to safeguard shareholders' interests.

Denis noted that the corporate governance literature deals mainly with the notions of ownership and control in various modern forms of business organisation.[24] So, corporate governance initially focused on the importance of organisational structures,[25] organisational hierarchy[26] and organisational decision-making[27] in regulating what people do in business organisations to stay accountable and understand how organisational actors deal with conflicts of interest across various stakeholders of the organisation.[28] Early scholarly work on corporate governance attempted to frame this by examining human agency using transaction cost economics approaches;[29] this is galvanised in the formulation of business contracts aimed at regulating behaviour. However, as Hart argued, the 'agency problem, or conflict of interest, involving members of the organisation [owners, managers, workers or consumers...] cannot be dealt with through a contract'.[30] Besides, formal contracts are forever partial due to the high transaction costs involved in writing comprehensive contracts on the one hand, and the fact that contracts are often subjected to human interpretation that could reduce the predictability of outcomes on the other.[31]

In an increasingly complex and networked world, however, it is accepted that formal institutions and mechanisms are limited in providing a full picture of what happens in reality, mainly because of the informal practices that predominate.[32] Much literature on corporate governance has stressed on formal mechanisms using economic and legal instruments to control the behaviour of firms. Yet there is now considerable recognition that informal practices matter. Scholars have long acknowledged that formal knowledge is often incomplete,[33] and that tacit forms of organisational knowledge developed through shared cultural norms[34] often escapes more explicit forms of governance structures. Furthermore, formal mechanisms of control are less effective in regulating modern organisations, which are characterised by the blurring of organisational boundaries and a greater emphasis on articulating stakeholder perspectives.[35] This, in turn, raises the need to consider interorganisational cooperation and a better understanding and mobilisation of social capital and trust between stakeholders.[36]

Another criticism of the corporate governance literature is that it focuses a lot of attention on the relationship between those who invest and senior management of firms, often with little regard for those further down the organisational hierarchy. Child and Rodrigues[37] bemoaned the unitarist view[38] that dominates the literature, and suggested that governance structures need to consider the implications of decentralisation of power and authority to the grassroots where conventional controls of hierarchy and Weberian bureaucracy become less appropriate. Furthermore, Child and Rodrigues argued that the assumption that 'top management has the means to ensure that a firm's operations are aligned with its strategic objectives' may be flawed because of 'the ability of middle managers and employees to sustain informal action that can distort management intentions'.[39] Contemporarily, this is reflected in growing interest in the practice-turn literature,[40] which subscribes to a more bottom-up approach that considers how informal practices can contribute to growth in organisations.

Conventional understanding of corporate governance has narrowly focused on sustaining shareholder interest.[41] The inclusion of practice-based literature and the pluralistic view suggested by Child and Rodrigues[42] to consider how actions taken by business organisations can have impacts much wider than to shareholders is pertinent to fairly recent agenda of corporate social responsibility. The study of corporate governance has indeed received renewed interest in part due to scandals and crises, for example, the collapse of large corporations like Enron, and the economic crises of Asia and Latin America in the 1990s.[43] Such scandals and crises have had major impacts on the general public, and corporate wrong-doing has been damaging for the public's perception about the world of business. Branston and colleagues observed that 'corporate governance is primarily concerned with the relationship between shareholders and managers/directors', but strongly believed that 'there is a need for accountability to all of the public who have an interest in the activities of a corporation'.[44] Aguilera and colleagues[45] contended that corporate governance should not only have the narrow focus on short-term financial returns, but also extend its scope to include longer term social and environmental responsibilities.

At the basic level, corporate social responsibility (CSR) is defined as 'the ethical behaviour of a company towards a society [...] management acting responsibly in its relationships with other stakeholders who have a legitimate interest in the business [...] the continuing commitment by business to behave ethically and contribute to economic development while improving the quality of life of the workforce and their families as well as of the local community and society at large'.[46] However, the enactment of corporate social responsibility is not without its problems. There are certainly fervent arguments made by opponents to corporate social responsibility to indicate that the existence of business organisations is solely for the purpose of making money for its shareholders.[47] Furthermore, there is still considerable debate on the empirical costs and benefits of CSR. McWilliams and colleagues suggested that 'CSR has been used as a synonym for business ethics, defined as tantamount to corporate philanthropy, and considered strictly as relating to environmental policy. CSR has also been confused with corporate social performance and corporate citizenship'.[48]

Despite the confusion, there is certainly a shift in corporate governance thinking to embrace the wider remit of businesses giving something back to the wider community. In this section, we have traced the evolution of corporate governance thinking from scholarly efforts to explain the relationship between investors and senior management of firms through better comprehension of organisational structures and formal control mechanisms (e.g. contracts), to the consideration of how senior managers interact with employees further down the organisational hierarchy to shape corporate actions, to the current focus on social responsibility. To a degree, efforts on CSR can be seen as a way to balance the profit-making motive that typifies the capitalistic mode of production. This is certainly timely given developments in the global financial crisis and the questions raised regarding the greed of bankers. The debate on social responsibility is likely to intensify as communities become empowered to demand accountability from business organisations for the corporate actions taken. In the next section, we review developments in the literature on community governance.

Community governance: revisiting social capital

We have established above that there are limits to the political and corporate levels of governance in solving the problems of society. So, whereas the role of government in policy-making is still crucial, and although the private sector is still important in wealth generation, the ability of governments to exert direct control and the narrow focus of corporations to sustain profits for the benefit of shareholders are no longer sufficient in contemporary society. There is an increasing role for communities to govern themselves. According to Bowles and Gintis, 'Communities are part of good governance because they address certain problems that cannot be handled either by individuals acting alone or by markets and governments'.[49]

In the bestseller *The Tipping Point*,[50] Malcolm Gladwell wrote about how the once-floundering brand *Hush Puppies* became fashionable because the bohemian community in Soho in New York started buying and wearing them. Similarly, Gladwell described how New York City Council tackled and effectively suppressed the spiralling rates of syphilis infection among drug users by distributing clean needles to drug users through the very community associated with the drug scene. Bowles and Gintis also described how Chicago residents worked together to reduce anti-social behaviour and crime among youngsters, and how Japanese fishermen in the Toyama Bay formed a cooperative to pool resources together to mitigate against problems of variable catch.[51] All these examples demonstrate the power that communities can have to achieve something positive. For Putnam, the mobilisation of community spirit brings about benefits associated with social capital, 'My wife and I have the good fortune to live in a neighbourhood of Cambridge, Massachusetts, that has a good deal of social capital: barbecues and cocktail parties and so on. I am able to be in Uppsala, Sweden, confident that my home is being protected by all that social capital, even though [...] I actually never go to the barbecues and cocktail parties [...] In the language of economics, social networks often have powerful externalities'.[52]

Throughout this book, we have emphasised the importance of networks, whether talking about how networks contributed to the formative development of leaders in Chapter 2, or how networks of government, corporations and community actors are critical in pulling together the various strands of sustainable development in Chapter 3, or even how our leading figures consider the need to harness more effective collaborations in the networked globalised world we live in today, networks have come to influence governance at all levels. Indeed, we have also argued for a greater role to be played by community actors in shaping a sustainable future. Bowles and Gintis postulated: 'community governance appears likely to assume more rather than less importance in the future. The reason is that the types of problems that communities solve, and which resist governmental and market solutions, arise when individuals interact in ways that cannot be regulated [...] due to the complexity of the interactions or the private or unverifiable nature of the information concerning the relevant transactions'.[53]

However, community governance is not unproblematic. One of the fundamental difficulties with community governance is to determine who is actually in control, or how much power communities really have. Thus, one could interpret the example of tackling syphilis among drug users in New York City as the local government taking control of choosing to work with the drug-taking community in the first place. Sullivan[54] suggested that the dilemma of community governance is twofold. On the one hand, it is not easy to determine whether a bottom-up approach is necessarily better or worse than a top-down approach; often the boundaries between both are not so distinct. On the other hand, notwithstanding the aspirations of empowering communities to take control of how their future is shaped, there is often the underlying problem of seeking appropriate representation of what communities need. As Sullivan noted, 'It is now taken for granted that representative democracy without participative democracy is insufficient [...] what is unclear in discussions about the respective contributions is how the two combine and what the balance is between'.[55] She suggested that good community governance should be built on a blend of three key strands, as follows:

- *Community government* where elected local government is fundamental to the system of community governance;
- *Local governance* where elected local government is one of many important actors at the local level and a successful system of community governance will be based on the most appropriate configuration of these actors rather than the privileging of the role of elected local government; and
- *Citizen governance* where citizens can be empowered using communitarian principles and correspondingly limit the power of elected local government.

Somerville questioned the adequacy of representation at the community level, as he noted, 'It is one thing to demonstrate that democratic community governance is desirable, but it is quite another to explain why it is so difficult to achieve'. He added that effective representation poses a challenge because in most societies, the power is still retained by a small group of ruling elite. Indeed, Somerville argued, 'elections can serve to strengthen the rule of an elite by conferring on it added legitimacy (it becomes known, for example, as 'the Establishment')', which, in turn, 'tend to build their agendas around purposes that have priority for the rich and powerful, because it is the latter who can contribute more resources to achieve purposes of any kind'.[56] A corollary of this is political apathy, which only serves to threaten the foundation of effective community governance.

Furthermore, how do we set community boundaries in an increasingly globalised world? Early on, we alluded to the paradoxical phenomenon of globalisation and how the rise of community governance is often a response by communities to assert their identity amidst competition in the face of globally mobile capital.[57] In the UK, as Mulgan[58] observed, there is a general perception that control of power is often lost to decision-makers in Europe. It would seem that the popularity of community governance is a direct response to the increasingly boundary-less world we live in today, such that the creation of a somewhat more local identity matters more. To further lend support to this, Putnam found that the increased trend of immigration has resulted in reduced social solidarity in ethnically diverse neighbourhoods in the USA, suggesting that the apparent dissolution of geographical borders resulting from intensification of migration has ironically resulted in the reinstatement of ethnic identities.[59]

So, in this section, we have seen the power of communities that can be mobilised to accrue benefits for society, whether this is through reduction of crime rates, or the promotion of health messages, or the marketisation of a product. There is indeed growing interest in the way communities can govern themselves to solve societal problems. However, a deeper review of the literature reveals that the notion of community governance is still in its infancy. Beneath the rhetoric of community governance lie the challenges of negotiating a balance between top-down and bottom-up approaches, and the problems associated with optimising effective representation. Furthermore, questions were also raised as to whether the growth of community governance is a direct response to the dismantling of boundaries in the networked globalised world we live in today. Nonetheless, just as Sobel[60] and others have observed, traditional forms of governance in the digital age are transforming. We have discussed how power might be shifting away from the state to the private sector to the community, and that new forms of authority and control need to be comprehended. As it stands, no one level of governance is adequate to explain how actors in government, corporations and communities can come together to engender change. In the next section, we consider the need for joined-up thinking in governance.

The need for joined-up governance

We have examined hitherto the various levels of governance in general. In summary, we have attempted to trace the shifts away from state-centred governance to the rising importance of the private sector and community to tackle the issues of society. No one level of governance is completely foolproof. Bowles and Gintis see a role for joined-up governance: '[...] well-working communities require a legal and governmental environment favourable to their functioning. [...] The face-to-face local interactions of community are thus not a substitute for effective government but rather a complement'.[61] Dunleavy and colleagues,[62] when undertaking a cross-national investigation of how IT was transforming political governance and citizen's participation, also concluded that digital technologies stand to empower individual citizens to autonomously engage in solving social problems, and that the role of the government is to facilitate this process.

Indeed, Geoff Mulgan, who coined the phrase 'joined-up thinking', asserted that joined-up governance is critical because the complexity of societal problems implies that the state is often limited in its power and resources to afford effective solutions without cooperation with the private sector and communities (see Boxes 4.2 and 4.3 for a couple of case examples). According to Mulgan,[63] a number of reasons explain the need for joined-up governance:

- *Complexity of problems*: contemporary issues such as poverty and competitiveness, family and environment warrant greater collaboration with businesses and the public, as solutions to such complex problems go beyond the ability or capability of traditional, singular government departments;
- *Limits of previous reform agendas*: experience has demonstrated the limitations of what governments could achieve on their own to combat societal problems;
- *Growing evidence of interconnectedness of problems*: there is growing evidence that societal problems are not mutually exclusive and that their solutions can be effectively found in collective action by various stakeholders;
- *Advances in technological and organisational techniques*: such advances make problem-solving more rapid and stimulate demand for problem-solving by networking;[64]
- *Influence of consumerism*: the rise of consumer sovereignty implies the need for government, private sector and community groups to embrace individuals at the grassroots level; and

Box 4.2. Building Futures East: helping the unemployed in the community gain skills

Background

Construction sectors across the world have been experiencing pressures of skills shortages.[65] In the UK, attempts have been made to remedy the situation using a number of supply-led initiatives, including inter alia the roll-out of a Construction Skills Certification Scheme, encouraging new entrants into the industry through outreach programmes in schools, and improving the public image of the industry through publicity campaigns. Construction employers have also adopted recruitment strategies that target non-traditional segments of the population (e.g. women and ethnic minorities), and designers have attempted to integrate more use of off-site prefabrication.

There is more that needs to be done, and the UK government seeks to raise the commitment (and efforts) of employers to engage in skills development practices. The Leitch review[66] of skills in England and Wales recommended more employers' input into training and education of the workforce. However, the literature has often criticised employers for their reluctance in engaging in skills development.[67] Often, coordination of skills training in Britain can be highly confusing for employers, given the myriad of organisations involved in skills policy formulation and delivery.[68] On the one hand, the UK government intends to get employers to be more proactive in training the workforce; on the other hand, employers find the political landscape for skills development cumbersome in terms of who is involved and funding mechanisms.

Building Futures East (BFE) is an example of how the gap between the government's intent and the employers' experience may be bridged for the benefit of a deprived community. Building Futures East was conceived in 2006 when Anthony Woods-Waters (CEO) and founder Rev. Fr. Michael Conaty wanted to do something to help the unemployed people in the deprived ward of Walker in the East End of Newcastle upon Tyne. As Rev. Fr. Michael Conaty had served in the local parish for a considerable number of years, he understood the nuances of the local community. Michael felt that the youngsters who are unemployed in Walker remained unemployed because of two main reasons: first, they were brought up in families who were also on long-term unemployment; and, second, the youngsters did not fit in with the academic route of the education system. For Michael and Anthony, the solution seemed straightforward: equip the youngsters with craft skills of building trades in a location near to their homes to prevent truancy that is not conventionally classroom-based.

Key challenges:

1. *Money:* Funding was sought from the government (through agencies such as the Learning and Skills Council), but was initially turned down. However, with perseverance, Anthony, who then worked for the local authority, found a

funding route through the European Social Fund, and secured money to fund the project as part of a regeneration bid.

2. *Private sector buy-in*: The intent of BFE is to equip the youngsters with craft skills and to instil in them a positive work ethic, in the hope of allowing them to obtain gainful employment. For this to work, employers needed to be integrated in the process. Anthony and Michael knew that major employers already had well-established recruitment strategies. Furthermore, most major employers are hollowed-out firms that do not necessarily employ trades people. They needed to get the small and medium-sized enterprises (SMEs) on board. With the funding in place, they manage to persuade SMEs in the local area to provide work placements while the youngsters were training, and the SMEs contributed £1 per working hour to the youngsters. Furthermore, a number of medium-sized firms were also altruistic enough to provide a shed in the local area for the training for virtually no financial reward.

3. *Community buy-in*: Long-term unemployment can be a sticky problem! For BFE to work, Anthony and Michael made sure that families were involved. At times, this meant that they had to knock on doors. Apart from inviting families to the training premises to see how future generations can work to get out of unemployment and poverty, Anthony also got the families to come in during lunchtime to help out in the canteen. This meant that they did not have to rely on external caterers, and gave the families a sense of enriching the experience of the youngsters.

Arguably, this story is filled with examples of how the local community and private companies have information about the community that does not necessarily get appreciated in government. In the process of forming BFE, however, the government (then represented by Anthony, and the European Social Fund), the local employers and the community were setting rules for making things happen for the unemployed, unskilled youngsters. Although its success remains to be seen and validated, BFE is an example of how joined-up governance could be powerful in solving the sticky issue of unemployment in society.

Box 4.3.   A Case of Second Health: mapping the patient and healthcare journey using Second Life

Background
Healthcare in Britain is transforming. The Department of Health (DoH) and the National Health Service (NHS) are modernising their thinking in terms of how healthcare is being delivered to the patient. Among other things, there is the strategic intent of bringing healthcare to the home and empowering patients to take prevention seriously. Another key development is in the field of primary healthcare. The DoH and NHS envisions are providing more 'one-stop-shop'

primary healthcare centres to enhance patient convenience. As part of this strategy, there is the intention to build polyclinics right across the country. Notwithstanding the currency of this contentious subject, Second Health (www.secondhealth.org) is at the forefront of designing these polyclinics.

Key challenges

1. *Not knowing the operational models of the future*: Tzortzopoulos and colleagues[69] reported on how an early scheme of modernising primary healthcare failed because the NHS was heavily involved in designing and constructing facilities before knowing how the healthcare professionals were going to work together in the future. Indeed, healthcare professionals in the primary care sector were used to working solely within their own specialisms. Sharing a building with other professionals (and other stakeholders e.g. local authorities) within a 'one-stop-shop' concept can prove demanding in terms of negotiating, for example, spatial requirements necessary to deliver the new mode of healthcare.

2. *The patient journey*: For the patient, the journey to and within a healthcare facility can impact on their satisfaction with the process. Research has been undertaken to look at how spatial design can impact on health outcomes as well.[70] However, visualising this journey, and optimising the journeys to suit everybody can be especially challenging.

Second Health is concerned with designing and visualising the future polyclinic using advanced gaming technology known as Second Life. They have piloted a few designs for the Central London area, where they get both healthcare professionals and the public to play with Second Life. This allows them to create their personal avatar (virtual identity) in order to create the space they hope for. This has enabled participants to negotiate spaces between them in the virtual world, thereby permitting communication between healthcare professionals and the potential patients from the local community that they serve. Designing and visualising in the virtual world also means that the process is less costly, when compared to the employment of professionals at the front-end of a traditional construction project.

- *Shift away from technical rationalism and dominance of atomisation of policy-making and implementation*: there is greater recognition of the weakness of technical rationalism and Weberian bureaucracy, and preference for more holistic thinking in society.

Table 4.1[71] summarises the key governance principles and measures with respect to our examination of political, corporate and community governance so far. In the next section, we explore how some of the key messages found in the mainstream governance literature are applicable in the context of the construction management literature.

Table 4.1 Measures of governance and paradigmatic shifts in the three levels of political, corporate and community governance

Principles of good governance	Governance measure	Political (state) level	Corporate level	Community level
Legitimacy and voice	Voice and accountability	Shift away from 'Government knows best' to encouragement of partnerships with the private sector and communities	Shift away from unitarist view of organisations to embracing pluralism, demonstrated through, for example, worker involvement initiatives	Shift towards greater inclusion and participation of community members through, for example, user consultation initiatives
Direction	Political strategy	Shift away from direct control of society's resources to a capacity to influence the outcomes in society	Shift towards the private sector putting more resources into the provision of infrastructure for society	Shift towards greater collective action in defining community action
Performance	Effectiveness of governance	Shift away from singular measures performance to responsiveness to the needs of society and setting of pluralistic targets of efficiency and effectiveness	Shift away from narrowly measuring short-term financial performance to include more qualitative measures to encourage learning and socially responsible behaviour	Shift towards more qualitative measures of well being that extend beyond just financial measures
Accountability	Control of corruption Regulatory burden	Shift away from departmental accountability to managers of public administration to greater transparency to the general public	Shift towards greater transparency and public accountability, as well as performance-based regulatory frameworks	Shift towards empowering citizens to hold their representatives accountable at all levels
Fairness	Rule of law	Shift towards more socially inclusive legislation as witnessed in, for example, the area of equality and diversity	Shift towards more socially responsible behaviour, and encouraging greater respect for employees	Shift towards widening participation by all segments of society

Governance in construction: the trends of privatisation and community engagement

We have outlined the paradigm shifts within the mainstream governance literature, which identified the changing emphasis from the state (political governance) to the private sector (corporate governance) to the grassroots (community governance). Arguably, this trend appears to manifest itself within the construction management literature. As suggested by our leading thinkers, one can observe that governments are moving away from direct provision of physical infrastructure to a role of facilitating more private sector involvement through procurement initiatives such as PFI and PPP. Alongside this, the structure of the construction firm is changing from one that directly employs craft workers within a traditional major contracting firm model, to a model of a hollowed-out firm[72] where the focus is mainly on managerial aspects of construction and major companies are heavily reliant on subcontracting and a casualised workforce. Alongside this, there is an increasing impetus for greater social and environmental responsibility placed on firms in the construction sector, with the growth of corporate social responsibility through initiatives like *Respect for People* and the *Considerate Constructors Scheme*, and calls for addressing climate change and the environmental agenda through such regulatory instruments as building regulations. Besides, the power of the consumer is ever more significant and the industry has to refine its user consultation procedures to ensure the satisfaction of not just the paying client, but also of all end-users of the built environment.

In this section, we review the construction management literature to distil out a number of salient issues relating to governance across the three levels of political governance, corporate governance and community governance. Specifically, we discuss the relationship between government and construction, the changing landscape of professionalism, the structure of the industry, and the importance of human relations. In so doing, we stress that any understanding of the construction industry needs to be set within the context of collaborations forged between government, corporate and community actors, thereby confirming the paradigm shifts in the mainstream governance literature. We will also utilise this review to compare and contrast the views raised by our interviewees in the section on governance of the industry above, so that overlaps and gaps for research and practice can be further galvanised.

The relationship between government and construction

We have seen thus far that the government is intimately connected with the affairs of the construction industry. Besides, in many countries, public sector work accounts for around half of national construction output. In this section, we review the relationship between what government does and the construction industry. We discuss this in relation to three critical areas. First, we explain how government agendas are changing and how the representation of construction in public policy is now enmeshed in the wider context of sustainable development, specifically in relation to the agendas of sustainable communities, energy use and climate change. We also explain how the government has moved away from direct provision of public infrastructure and how such an arm's length approach to enabling the delivery process is a symptom of depoliticisation of political governance, where governments retain control yet devolve responsibility to external parties such as the private sector. We also discuss the efficacy of the government's role in regulating private sector affairs, and observe that, in some cases, this is restricted by the law of unintended consequences and the complexities surrounding globalisation.

Changing government agendas reflected in representation of construction in government

Murray and Langford[73] argued that direct control by the government over the construction sector has diminished over the last 50 years. Taking the UK as an example, they observed that whereas a government official used to have a clearly defined remit to oversee the construction industry, this role is increasingly becoming marginalised. Indeed, Murray and Langford found a lack of clarity as they struggled to track a specific, named official who acted in some capacity as a 'Minister for Construction' in recent times. They argued that this creates a representational vacuum, which, in turn, reduces the ability of the industry to effectively lobby their interests in the political domain.

Of course, as we alluded to above, there is also an argument that the societal problems of today have become so much more complex that a single 'Department for Public Construction' (or 'Public Works' as it used to be called) would be deemed inadequate. Indeed, the contemporary pursuit of sustainable development and concerns surrounding climate change, energy consumption and security all require interdepartmental cooperation, often including cross-national coordination. Thus, the intersections of these new agendas are often reflected in new names adopted by government departments, e.g. 'Department for Communities and Local Government' or 'Department of Energy and Climate Change', which have associations with the work of the construction industry.

As discussed earlier, what underpins the reconfiguration of government departments is a political ideological change, which is in line with the shift of responsibility in provision of public sector construction towards the state's role in facilitating the process of delivery. This is demonstrated, for instance, in the decline of direct employment of craft workers in local government, a point we will return to when we discuss the structure of the industry below. Consequently, it is observed that the order of the day seems to be dominated by an agenda of encouraging competition, which, in turn, necessitates the strengthening of collaboration with the private sector. This is also within the context of a deeper alignment at an international level. So, the rebranding of UK government departments can be seen as part of wider efforts to connect their remits with the European counterparts in an era of more intense global cooperation on socio-economic matters.

Depoliticisation of the state in greater engagement with the private sector

A number of reasons account for the changes presented in the preceding subsection. For Kerr, the changes in political ideology and practice are inevitable, as he provided three reasons to explain increasing private sector involvement in public procurement of construction projects: 'The first is the requirement to transfer "appropriate" risk to the private sector [...] Secondly, through the requirement to define risk and to quantify future life-cycle costs and future service needs, the PFI is attempting to force greater objectification and "marketisation" into the provision of "public" services. Thirdly, through displacing the service provision labour process from the public to the private sector, the PFI is attempting to depoliticise the "labour problem" by making it the responsibility of the private sector'.[74]

Kerr maintained that financial risks associated with capital projects, especially amidst tighter fiscal conditions since the 1970s, remained the *raison d'être* for increasing use of private sector finance in delivering public sector construction. It certainly makes sense for the state to maintain an arm's length approach when procuring construction work in order to '[. . .] increase the flexibility of the state to be able to change the type, standard and quantity of services provided and their spatial manifestation through the built environment [and to] help massage the Public Sector Borrowing Requirement (PSBR)'.[75]

In a similar vein, Lansley traced the developments that led to the rise of the private sector involvement in financing public sector construction. He ascribed this to the challenges faced by the state of sustaining steady streams of work with increasing lack of funds: 'The 1970s witnessed a dramatic decline in the level of government funding for construction work. The decline was partly due to lack of funds but it was also a consequence of supply having met demand in the late 1960s. In the early 1970s there were no longer outright shortages of particular types of building or of physical infrastructure. Rather, investment priorities became subject to political interpretations of need and to political lobbies, a process which the industry, having had no previous experience, found difficult to manage [. . .] Not only are political interpretations of need giving way to commercial interpretations, but those companies wishing to have a stake in such projects are becoming more fully involved both in financial engineering and in public and community relations activities'.[76] These explanations certainly resonate with the views of our leading thinkers, especially those of Sir Michael Latham and Nick Raynsford, presented above.

Of course, the definition of risks and what constitutes appropriate levels of risk transfer to the private sector remain contentious issues that continue to be debated in public discourse. Nonetheless, the changing nature of the relationship between the public and private sectors over the last two decades through, for example, PFI and various configurations of PPP, represents the constant transformation of the process of privatisation of public sector construction. Arguably, the transfer of responsibility of provision to the private sector has ensured that economic and financial factors prevail in the understanding of risk, and this, in turn, diminishes the social construct that conventionally attached to public provision of infrastructure development. Such a transformation necessitates the refinement of rules associated with procurement, as collaborations between public sector bodies and private sector providers continually undergo a process of learning,[77] and configurations of integration between public and private entities remain in constant flux as new forms of procurement[78] are introduced.

One of the major consequences of a greater role played by the private sector in delivering public sector construction is the depoliticisation[79] of the building process. There is the removal of what Lansley[80] termed the political nature of the decision-making process. For Burnham, 'State managers retain, in many instances, arm's length control over crucial economic and social processes, whilst simultaneously benefiting from the distancing effects of depoliticisation. As a form of politics it seeks to change market expectations regarding the effectiveness and credibility of policy-making in addition to shielding the government from the consequences of unpopular policies'.[81] Accordingly, depoliticisation as a governing strategy takes three forms, including:

- Reassignment of tasks away from the party in office to a number of ostensibly 'non-political' bodies;
- Increase the accountability, transparency and external validation of policy; and
- Adoption of binding credible 'rules'.

Burnham's observations clearly apply to the construction sector. In the UK, for instance, we have seen the reassignment of responsibility for social housing to a range of providers including registered social landlords (RSLs) and private housing associations, regulated by a 'non-political' government agency known as The Housing Corporation (now subsumed under the Homes and Communities Agency). Elsewhere, we have seen a surge in the privatisation of public space,[82] which has given rise to the phenomenon of 'gated communities' in the USA where the private sector undertakes the enforcement of street security, a responsibility that has traditionally been within the realm of state control. At the same time, the formation of organisations like the Commission for Architecture and the Built Environment (CABE) aimed at promoting design and architecture to raise the standard of the built environment, particularly in healthcare and education, is an example of increasing accountability, transparency and external validation of public policy. Furthermore, recent changes to Part L of the Building Regulations that govern energy efficiency of buildings stem largely from binding credible 'rules' derived from the EU legislative machine.

   The implication of the process of depoliticisation of public sector construction is the emergence of greater pluralism of the construction industry. On the one hand, such pluralism is welcome given the diverse needs of various segments of society that have to be met through provision of the built environment. On the other hand, however, if managed poorly, such pluralism can lead to fragmentation of the sector that can seem confusing to the lay stakeholder. However, government intervention through legislation can be a plausible tactic to address this confusion. And, as the preceding section on political governance suggests, the state given the will still retains control over the power to legislate.

Government's role in regulating private sector affairs

According to Barry, 'In politics the collective is not a given, but an entity in process. The fact that there is never likely to be a consensus about what the collective is and what individual rights and duties are does not prevent the emergence of a common view. Conversely, the need for a common view does not make the fact of disagreement evaporate. Instead it means that our basis for common action in matters of justice has to be forged in the heat of our disagreements. In general, legislation and technical regulation have the effects of placing actions and objects (provisionally) outside the realm of public contestation, thereby regularizing the conduct of economic and social life, with both beneficial and negative consequences'.[83] Put another way, there is greater recognition that the power of the government in terms of their ability to legislate remains in constant flux of change. Thus, it would seem that what really matters most is emergent lessons about the nature of change and its implications are constantly being captured and disseminated. The structures of governance become secondary, as what is significant is the understanding that the state of becoming is equally, if not more, important than the state of being itself.

One area where there is truism in this is the realm of government legislation. There is a growing acceptance of the limitations of mechanistic rules of engagement where legislation is concerned. Each government rule of law is subject to unintended consequences, which result in gaps of regulating behaviour. An example of market regulation in the Spanish construction industry exemplifies this. González observed that 'more restrictive labor and tax regulations have induced parties to substitute market contracts for labor contracts because of the need to avoid moral hazards' and argued '[. . .] that this explains the increased fragmentation of the Spanish construction industry'.[84] Indeed, this echoes with the experiences of construction sectors based in the developed world where more protectionist labour regulations exist, and where the response of companies in the construction industry is to intensify multi-level subcontracting in order to shed economic risks and their associated financial costs. So, an attempt by the Spanish government to regulate the construction labour market led to a sizeable alteration to the structure of the industry. Protectionist labour markets were only partial in the regulation of the industry. Subsequently, the Spanish Ministry of Work and Social Affairs have introduced legislation to curtail the practice of subcontracting specifically within the construction sector, limiting any supply chain arrangements to a maximum of three tiers.[85]

Of course, the effects of this legislation in terms of provisionally engendering change in the behaviour of the private sector remain to be seen. But, if the example can prove that the Spanish government has found an antidote to the perpetuation of flexible labour markets, its transference, at least at a European level, might become a material development in the short term. Nonetheless, the effect of government legislation on altering behaviour of the industry is significant. Moreover, we acknowledge the complexities of government legislation and how any regulatory effort can only be partial in attaining the intended outcomes in reality. Constant monitoring and adaptation of the regulatory function of government is, therefore, most critical.

Another complexity of the government's role that needs to be discussed is the effectiveness of intranational response within an increasingly globalised context. In Chapter 3, we called for an institutionally coordinated response to achieving sustainable development. The role governments across the world can play to strengthen public institutions for the betterment of the construction sector is undoubtedly crucial. However, the practice here is not homogenous across countries in the world, and this can be problematic. Take the issue of human capital for example. Studies abound that suggest strong institutional capacity aimed at developing the workforce translates into a more stable skills base, which, in turn, increases productive capacity of the economy.[86] Yet, the way different countries approach this varies and the trend of global mobility of capital and labour stands to threaten intranational efforts to coordinate an effective response to the skills development agenda. The Spanish case of regulating labour markets described above lends further support to this assertion.

The issue of labour migration serves to explain how the capacity of national governments to design effective regulations nationally can be diminished within a cross-national context. From developed countries such as the USA and the UK to developing nations like the United Arab Emirates, there is a growing reliance on migrant workers to undertake construction work. Interestingly, tensions between indigenous and migrant worker communities, challenges with cross-cultural working resulting in communication problems and health and safety risks, and instances of abuse and exploitation of migrant workers have been reported in countries that are dependent on migrant workers to deliver construction projects. Yet, Bartram observed that state intervention through labour policies and the strength of institutions such as trade unions, although useful in alleviating some of the problems associated with migrant worker employment, can be divergent across different countries.[87] He found that there are countries (e.g. Japan) where the state regulates 'guestworker' immigration heavily in relation to wider socio-economic and political concerns, whereas other countries (e.g. Israel) have got weaker institutional structures that result in short-term benefits accrued to employers. Consequently, there is the fear that mobility of capital might mean that employers can exercise the choice to relocate to countries where there are weaker regulatory regimes where labour markets are concerned. Indeed, Chan and colleagues argued, when discussing the regulation of migrant worker employment in the UK construction industry, that national migration policies themselves would fail if wider political and economic considerations for industrial policy within a European or even international perspective are not accounted for.[88]

We have consistently seen the role of the government transforming from that of being direct provider of public infrastructure development to that of enabling its delivery process. In this section, we discussed how the devolution of responsibility of provision to the private sector is inevitable given the inward pressures of smaller fiscal budgets and outward pressures of globalisation. We have also ascertained that the shifting trends of *governance without government*, as explained above, apply to the context of construction, where the power of governments to influence the industry through legislation remains intact, but yet control over the delivery process is relinquished to the private sector. This process is described as the depoliticisation of the state in shaping the affairs of the industry at large. We also examined how the structure of the government has changed such that direct representation of the industry in the political domain has dissolved with the genesis of complex societal problems of sustainable development. The quest to develop sustainable communities and the tackling of climate change and energy use imply a need for intersecting departmental efforts, often within an international context. Notwithstanding the power of governments to retain control over regulatory regimes, we also discussed how this power can be restricted by the law of unintended consequences and divergence between countries. As a result, formal, mechanistic rules of engagement in political governance become less relevant, as we illustrated with examples from the regulation of construction labour markets. The structure of political governance and the construction industry is, therefore, less of a static concept, but more in a state of constant flux where the need to capture emergent lessons becomes critical. In the next section, we focus our attention on examining the structure of the industry from a private sector perspective.

Structure of the industry

Just as political structures become more fluid, the nature of companies operating in the construction industry demands further scrutiny. In this section, we review the literature pertaining to corporate governance in construction to assert that traditional organisational boundaries become somewhat meaningless. Early firm-level studies artificially put forward an understanding of organisations as coherent and consistent entities. The comprehension of how construction goods and services are delivered by conventional organisational functions is becoming outmoded, especially with the intensification of subcontracting and use of global supply chains. Consequently, there is a need to shine light on the relatively underexplored area of interorganisational relations. This section concludes with identification of a number of research and practical gaps aimed at further exploring how configurations of collaborations within the private sector can be shaped and the implications of such arrangements on power relations. Explaining these gaps will help resolve some of the ambiguities of control confronted by firms working together to design and construct the built environment.

Tracing the early roots of firm-level studies

Early scholars examining the nature of the construction industry classically studied the way construction firms are organised. At its heart, a well-established textbook like *Modern construction management*[89] systematically adopts a functional approach to discuss what goes on in the various departments of a typical construction company, including inter alia the management of finance and budgets, quality control, workforce motivation, production planning and plant management. Indeed, the importance of exerting management control along hierarchical corporate structures appears to dominate our somewhat Taylorist comprehension of construction, certainly in the latter half of the twentieth century. Thus, the economic theory of the firm has certainly underpinned the writings of leading thinkers in the field.[90]

Contemporary studies into corporate governance in construction have centred mostly on firm-level analyses, often investigating the influences of organisational structures on performance. For example, Rebeiz and Salameh[91] investigated the extent to which the configuration of the top 100 construction firms in the USA have an impact on their resultant financial performance; they concluded that construction firms that have a critical mass of outside independent directors and that dissociated the roles of CEO and Chairmanship of the board translate into superior financial market returns for the firms. In another study investigating corrupt practices in the UK construction industry, the CIOB surveyed industry practitioners on their perceptions of how pervasive corruption was in the industry and the extent to which their company's practices contributed to this phenomenon.[92] Implicit in the study is that the industry is merely the sum of its parts, and that firm-level analysis in itself would be adequate to explain what goes on in the industry in the macro context.

However, as mentioned above, the study of construction firms as coherent and consistent organisational entities is becoming irrelevant in the context of firms becoming 'hollowed-out' organisations, where the perpetuation of the flexible organisation implies the contracting-out of the activities of construction to ever more complex supply chains. In the following subsections, we delve into this trend in greater detail as we argue for the need to reset the focus of the critical nature of interorganisational dynamics that matter in the study of how construction firms operate.

Fragmentation or diversity: significance of interorganisational dynamics of construction

Pearce[93] acknowledged the pluralistic nature of the construction industry in terms of the make-up of firms that operate within the broad definition of construction, which integrates firms that are traditionally categorised in such other industrial classifications as manufacturing and services industry. Indeed, the peculiarity that differentiates the work undertaken by the broader definition of the construction industry from other industries is the way the project dominates our thinking about its production process.

The idea of construction[94] has seen a shift away from the narrow focus on the construction firm as the basis of economic activity to the notion of an industry made up of an agglomeration of multiple contexts and stakeholders. For Groák, 'different sectors of construction use fundamentally distinct resource and skill bases'; as such, he argued that it is no longer sufficient to promulgate 'the idea of "one technology, one industry"'.[95] He asserted that the inherent weakness in much research about the construction industry is the 'failure to recognise that the site was the defining locus of production organisation'[96] and that construction is 'essentially organised around the project, not the firm', recommending that any analytical framework should embrace the legitimately 'ad hoc' nature of construction projects as 'temporary coalitions in a turbulent environment requiring unpredictable (but inventable) configurations of supply industries and technical skills'.[97] Accordingly, this idiosyncrasy of construction demands alternative theoretical frameworks that can explain the organisation of construction work beyond firm-centric means. Studying corporate governance in construction should, therefore, transcend firm-level analyses to examine how networks of a myriad of firms (also known as the supply chain) come together to deliver construction projects.

This call was, of course, not new. In the Phelps-Brown inquiry into matters of organisation of labour in building and civil engineering, it was noted that 'the criticisms ranged at the fragmented nature of the industry arise from a lack of understanding of its function'.[98] Such inherent diversity in construction, as opposed to fragmentation[99] is what makes the construction industry so distinctive. Lord Charles Percy Snow, former Minister of Technology in the UK in the 1960s, famously remarked that what makes construction work unique is that whereas the line of work usually passes through the hands of people in most production systems, it is a line of people passing through work in the case of construction. The construction firm, therefore, cannot be an appropriate organisational form for analysing the nature of construction, as every construction project is organised as a quasi-firm.[100] More recently, Pryke offers this explanation: 'The construction industry appears to be evolving procurement and management systems that lie somewhere between the market and hierarchy models, with packages of work let, possibly, through a market driven approach, but subsequently managed in a hierarchical context within the environment of the *temporary project coalition* [. . .] It is suggested that we might regard the construction project as a network of firms working together for the purpose of a project (and, increasingly, for the purpose of a number of projects under a partnering arrangement).[. . .] There is, therefore, a need for a new way of identifying and describing the roles and relationships between the project actors within these project networks incorporating [. . .] non-hierarchical relationships between actors'.[101]

Inherent ambiguities of interorganisational relations: gaps for research and practice

The role firms in the construction industry play within civic society is evolving. The political trend of governance without government has seen increasing involvement of the private sector, and, in particular, construction organisations, in providing the infrastructure for services that were once conceived as the domain of the public sector. At the same time, the structures that govern the relationships forged between the supply networks of firms demand closer scrutiny. Firm-level analysis, especially in construction, is no longer sufficient to explain the diversity of what goes on in the industry.

The intensification of such trends as globalisation and outsourcing has resulted in the growth of subcontracting and the rising use of non-traditional forms of labour (e.g. self-employment and migrant workers), consequently leading to the emergence of new organisational forms of business; e.g. the permeable organisation,[102] the networked organisation[103] and the project-based organisation.[104] These new organisational forms, what Marchington and colleagues[105] term as disordering hierarchies, challenge traditional notions of power and control, and create a need to reconceptualise management practices, especially across firms. A critical case in point is the Olympic Development Authority (ODA) that is delivering the infrastructure for the London games in 2012; it is reported that this will see contracts issued to as many as 2000 firms in constructing the facilities.[106] Understanding governance of the network of organisations delivering such a colossal scheme requires not only an examination of how managerial efficiency is achieved, but also the power relationships and social capital created between the parties involved.

Rubery and colleagues argued that managing people across traditional firm boundaries inherently creates HR ambiguities in a number of areas; these arise due to tensions in specifying supervision and control regimes, disciplinary and grievance procedures, working conditions, legal and statutory obligations, loyalty and commitment, and worker representation.[107] Marchington and Vincent noted that the extent to which these ambiguities are clarified depends on the nature of collaboration between the firms within the supply chain, and whether these are governed through obligational contractual relationships or arm's length arrangements.[108] It is suggested that more work needs to be done to examine how alignment, integration and consistency in HRM across companies can be (or even should be) achieved.[109]

For Winch, governance in construction is much more than conventional corporate governance within a typical firm and includes what he considers to be 'an important factor in external horizontal governance'. He suggested that 'the shift from traditional subcontracting to supply chain management is first and foremost about the exercise of power

[...] Some clients are starting to use supply chain management principles to govern the project chain, thereby reducing reliance on complex contracts. [...] Where the parties do not know each other, the professional governance of transactions offers considerable advantages: its trust inducing properties generated by professional validation and grievance procedures, professional codes of ethics, and the central importance or reputation in supplier selection all favour the generation of trust between parties which do not know each other'.[110] This, as we shall see in the discussion on human relations below, requires methodological approaches that can dissect the interorganisational connections that are so vital in resolving how construction projects are governed.

Indeed, new tools are needed to solve old problems. The shift away from firm-centric analysis has orchestrated research interest in supply chain management and integration. Yet, beyond the basic emphasis on waste minimisation[111] and simple exhortations for more partnering and integration,[112] there remain further questions that have yet been resolved in our understanding of interorganisational dynamics in construction. London and Kenley[113] pose the following questions as future research agenda for understanding the governance of construction organisations in a supply chain context:

- What is the overall nature of the organisational relationships along the supply chain?
- What is the nature of the competitive environments within which organisations operate, and how will this affect the performance of firms in that market?
- How do firms source their suppliers?
- How does a supply chain form?
- Who actually supplies to whom?
- How is sourcing organised?
- What are the power relationships between firms and their suppliers along the chain?
- How do we analyse such fundamental structural and behavioural properties in the supply chain?

It is here that social network theory might offer some insights into the interorganisational dynamics that typify relationships in construction. Pryke,[114] drawing inspiration from Nohria and Eccles,[115] suggested that social network theory is extremely appropriate for the study of construction project organisations as the industry, as alluded to above, is made up of a multitude of interconnected firms working together to deliver projects. Accordingly, the intersections between the firms are complex and overlapping, and understanding any actions or behaviours of individual actors, and any comparative study of organisations, would need to account for the network characteristics.

Pryke found social network analysis appealing because it not only allows a more accurate and dynamic exposition of project structures and process that focuses on explaining governance in terms of 'networks of contractual relationships and networks of performance incentive relationships',[116] but also other researchers have employed social network analysis to examine the interactions between people in construction. Swan and colleagues, for instance, explored how trust manifests in construction in their quest to develop a trust inventory for construction, as they sought to identify key players to whom trust matters.[117] Attempts have also been made to see how innovation happens in the web of interpersonal relationships that evolve in a typical construction supply chain.[118] These techniques are beginning to examine the networked world we live in today in much greater detail, transcending basic firm-level investigations or dyadic, interfirm explorations featured in much early research on supply chains.[119] Nonetheless, there is more scope for engaging in more textured, interpretive analysis[120] of social dynamics beyond mere quantification of relationships offered by social network analyses.[121]

To summarise, our understanding of corporate governance in construction is becoming more sophisticated. There is growing acknowledgement that theoretical models of the firm are limited in explaining the particularities of how construction work is delivered. This is because the project-based nature of the construction industry implies a need to focus the attention on how firms, often transcending clear distinctions of industrial classifications, collaborate effectively to design and construct the built environment. Therefore, what matters more than the description of departmental functions of construction businesses is the need to explain the interorganisational dynamics that are crucial in the production of the built environment. It is likely that more research will be undertaken to distil out the critical challenges encountered by firms in engaging with more cooperative forms of business relationships. Besides, collaborations are often easier said than done; there are inherent ambiguities of control that make the study of power relationships in construction projects very pertinent. Moreover, knowledge on the rules of engagement in private sector collaborations within the construction sector still remains elusive. However, progress is made with the employment of such techniques as social network analysis. Of course, there is a social dimension that has not been addressed in the discussion of the structure of the industry. We now turn to examine the importance of society as an overarching concern for the operations of the industry in the next section, as we review the literature on the changing landscape of professionalism in the construction industry.

Changing landscape of professionalism

We are living in a highly mediatised world and so image matters a lot! In this section, we review how the construction industry can be negatively perceived in the public domain, through stereotypical views of the builder and the increasing disregard for the professional in society. We review the literature on the pitfalls of professionalism in the construction industry, focusing on the exclusive, privileged nature that serves to reinforce traditional class boundaries in society. Professionalism in the construction industry is, indeed, at a crossroads. To a large extent, the make-up of those working in the industry is changing; the time-served crafts worker who can work his/her way up the organisation is being replaced by the professional manager, often educated without the depth of technical knowledge of the past. We discuss how, in contemporary society, the increased professionalisation of the industry has paradoxically created a barrier to innovation and productive industrial development. In part, this is due to the crisis of the professions, characterised by the perpetuation of self-interest as opposed to serving the public good. Yet, in the era of social responsibility, and with the rise of the educated and discerning customer, the pressure is on for professionals to deliver. This section also discusses the future of professionalism in the construction industry, given these changing social dynamics.

Image of construction: reinforcing boundaries

The construction industry generally suffers from having a poor public image. Tradition-ally, construction work is deemed to be dirty and unpleasant, and the industry is plagued by a preconceived notion that it is unprofessional.[122] Popular media are often unkind to the construction fraternity as well, in that terms like 'cowboy builders', 'rogue traders' and the 'builder's bum' fuel derogatory views about the sector.[123] These characteristics corroborate with the Market and Opinion Research International (MORI) poll in 1998,[124] which found that 16-year-old school pupils perceive the construction industry as a low status, dirty and lowly paid industry in which to work. The poll also suggested that pupils who were attracted to construction as a career route chose the industry because they wanted to work with their hands, thereby connecting the industry with manual work.

Findings from the MORI poll have been rehearsed in a number of subsequent writings. For instance, Moore[125] commented that the industry fails to attract school pupils with high academic achievements; these students are often drawn to other sectors.[126] Langford and Robson also made a fascinating analysis of cinema representation of two built environment professions – the engineer and lawyer – and discovered that 'cinema values the product of engineers as part of the process of making films; the lives of engineers are of little interest; in contrast, the legal profession is valued as a centrepiece of "real life" dramas'.[127] Indeed, the limelight appears to shine on certain professions while casting shadows of doubt on others, all of whom in principle deliver the built environment. Another notorious example is the demarcation between design (the architect) and construction (the builder),[128] which serves only to aggrandise the much-criticised fragmentation of the industry.

Arguably, such polarisation does little to improve the image of a vocational sector like construction. Furthermore, there is a trend towards encouraging flatter organisations in the business world, which consequently strips out traditional career progression routes of working one's way up the organisation. Specifically in construction, the ability for a time-served tradesperson to progress up the managerial career ladder in organisations is increasingly being replaced by the recruitment of the professional manager often educated through the further/higher education system. This is illustrated in the British example, where the UK government's policy towards attainment of 50% participation rate of higher education[129] and lukewarm reception of the Tomlinson[130] report's recommendation for greater emphasis of vocational education reinforce this polarisation.[131] The attitude of public policy-makers in this respect signals a preference for academic skills over vocational coined in the rhetoric of reaping the benefits of a knowledge economy, all of which does little to boost the image of construction. To exacerbate the situation further, there is also erosion of the vocational skills content provided by the apprenticeship system, which the industry has until recently relied on for the provision of the skilled workforce.[132]

There are consequences of such an approach, as Clarke and Herrmann[133] observed how the institutional framework in Britain has served only to reinforce class-based divisions of labour, ultimately resulting in the fossilisation of the definition of skills in the industry, which, in turn, leads to the perpetuation of outmoded industrial practices. It is no wonder that the industry is often considered to be backward.[134] Steps have nonetheless been taken in an attempt to improve the image of the industry.[135] These range from popular culture with the fictional cartoon character 'Bob the Builder',[136] to more mainstream initiatives of engaging with non-traditional groups like women and ethnic minority groups,[137] although more work needs to be done to improve diversity in construction[138] and correct the problem of poor public image of the industry as a whole.

A professional crisis? how professionals engage in contemporary society

So, the time-served tradesperson who works his way up to become a manager with strong foundational knowledge and technical expertise is gradually being replaced by the professional manager. The professionalisation agenda has certainly come of age in the construction industry globally. Yet the landscape of professionalism is also undergoing a crisis of its own. Who would have thought, for instance, that the reputation of doctors would become tarnished by the acts of one Harold Shipman, a British general practitioner and convicted serial killer of more than 200 patients in 2000? How have bankers become viewed with disdain by the public as their imprudence led to one of the most serious global financial crises since the Great Depression of the 1930s? It would appear that a gulf of mistrust is developing between civic society and the very professions that have provided much ingenuity in developing industrialised nations in the nineteenth and twentieth centuries. Such mistrust poses the question as to what the essence of professionalism today really is and should be?

Apart from the issue of the unprofessional 'rogue trader' image that the construction industry portrays, professionalism is increasingly blamed for many of the alleged problems (e.g. inefficient practices, lack of innovation and modernisation) faced by the industry. Winch, tracing the development of professionalism in the sector, noted that, 'By the mid-1960s, the professional system had become 'the establishment' [...] more concerned with protecting its own interests than meeting client needs. It allocated roles, defined responsibilities, and specified liabilities'.[139] It would seem that professionals today are less concerned about generating good for society and more interested in maintaining their monopoly to practise as professionals.

For Clarke and Herrmann, it is the British form of professionalism that has become a barrier to development and continuous improvement in UK construction: 'professional institutions are incorporated on the ability to demonstrate an exclusive area of knowledge or skill, which remains fixed. Any new or overlapping areas of activity/knowledge, or areas falling between institutions, become issues of conflict, rivalry, demarcation or exclusion [...] The institutions have a vested interest in maintaining their monopoly and little incentive to cooperate or merge with other institutions. Ministerial accountability and the monitoring of the public interest are also weak and remote. Indeed, as the commercial interests of institutions have become increasingly important, so the aim of serving the public good is put into question'.[140] A corollary of this is the increased fragmentation of the professional landscape contributing to the adversarial relationships between parties delivering construction projects identified in numerous governmental inquiries and reports.[141]

Empirical evidence is stacking up in support of this assertion. In comparing the procurement and management arrangements of UK and French construction projects using social network analysis, Pryke and Pearson found that because 'The French construction project seems to be less professionalized than the typical UK project [. . .] The French appear to be able to build very quickly and cheaply',[142] even after allowing for problems faced in transnational comparisons. They concluded that the success of the French team observed might be attributed to the relative simplicity and integration in interprofessional arrangements, and 'the devolution of all detailed design to contractors and specialist subcontractors, perhaps coupled with the absence of the traditional UK architect's role'. They added that the speed of the French team was also helped by using 'a very transparent and simple, yet extensive, system of activity-related financial penalties', an incentive system that 'effectively require all the interdependent actors to perform their duties in a manner which enabled the project to progress [. . .] the culture is one of good business relationships and is quite hard nosed rather than the more idealistic culture advocated by Egan'.[143] This study demonstrated that the absence of a fragmented professional landscape, together with a financial incentive and contractual system that promoted collective working towards clear performance goals could result in less interprofessional coordination (and even interprofessional bickering) and greater efficiency gains.[144]

To arrest further fragmentation of professionalism, and in line with the trend towards more interdisciplinary working in the industry, calls have been made for greater regulation of the professions within a comprehensive industrial system. Clarke and Herrmann maintained that the trouble with the British system of professional institutions is that institutions are only loosely governed by The Privy Council, which grants institutions 'a privilege [. . .] in perpetuity, only revoked in the case of a serious breach [. . .] The institutions largely determine what their members learn [. . .] with the state playing no direct regulatory role'. Consequently, they argued, 'the role of universities in Britain in the education of built environment professionals is very restricted because of concerns about accreditation and the importance of professional membership to student careers',[145] thereby reinforcing a utilitarian approach to skills formation of built environment professionals in Britain that would impede the advancement of the industry. By contrast, Clarke and Herrmann examined the professional landscape in Germany and observed that institutional structures are much simpler and there is relatively less fragmentation and division of labour as compared with Britain.[146] This is the result of stronger involvement of the social partners (i.e. the state, employer and employee representation) and particularly due to tighter state regulation of the professions in terms of 'establishment, structuring, administration and training of professionals'.[147] So, it seems that professionalism – or at least the British form of professionalism characterised by deepening fragmentation resulting from a neo-liberalist approach to state regulation – has led to the maintenance of a rigidly traditional route to skills formation and a corresponding low performance.

Beyond the managerial agenda of professionalism in construction

A key feature of contemporary professionalism is how much of the industry's agenda is framed by the prevailing thought of managerial functionalism. To put it simply, many of the troubles confronted by the industry can be remedied through so-called cutting edge management thinking and the actions and interventions of professional managers. Furthermore, there is an obsession with the performance agenda, often defined narrowly by short-term targets. Consequently, this places much emphasis on mechanistic approaches to gain quick fixes in solving business problems and/or to achieve efficiency. Researchers and professionals have often embarked on an expedition to search for a golden nugget that can offer a universal panacea to the problems faced by the industry. Such an approach is flawed in three ways. First, such a focus can often be misplaced as the validity of the managerial problems identified can sometimes be questionable. Besides, problems associated with business corporations can be derived from wider socio-political considerations, and so, any intervention framed solely in the context of corporate-level analysis, and without proper engagement with government and community actors, can sometimes be futile. Second, managerial functionalism forces professionals to think only in the short term, and can distract them from developing longer term prospects for the industry. Third, the pursuit of managerial functionalism can result in a somewhat dismissive approach towards social issues, and lead to a disregard on doing good for the wider society.

The problems of managerial functionalism have been well documented in the construction management literature. Chan and Kaka,[148] for instance, observed that construction managers often put task-oriented, technical planning issues ahead of matters that affect wider social factors such as employment relations. Green[149] is often sceptical of the industry's keen interest in lean thinking, suggesting that managers often do not consider the impacts of such managerial tools and techniques on issues like workers' autonomy, job satisfaction and operational culture. Druker and colleagues[150] distinguished between hard and soft management as they suggested that the emphasis on construction planning meant that human resources are often treated as another factor of production and implied that construction organisations rarely operationalise the rhetoric of soft HRM, which considers worker welfare and longer term industrial development. In fact, many empirical studies support the assertion that managerial functionalism often meant that the human being and the dynamics of human interaction often features less prevalently in the studies on construction.[151] Indeed, the construction management literature is often littered with the diligent reporting of research findings into the latest best-practice model or the updated list of critical success factors, often emphasising a purely managerial perspective.

Progress is, however, slow in broadening research efforts to examine particular aspects beyond managerial functionalism. Early opponents of such a narrowly confined approach to defining the social problems and cures of construction argued that the prevailing positivistic framework relies heavily on the assumption that human beings are objectified individuals, and that the rationalistic principles have become institutionalised and deeply embedded within the construction management research and even practice community. Often reduced in the understanding of a singular culture of the industry, Seymour and Rooke objected to the dominant discourse of construction management led by those who '[take] for granted the interpretative frameworks that are used to organise and communicate perception, thus effectively ignoring them. Instead of investigating the interpretations of others, they simply assert one of their own'.[152]

The ills of construction are far too often pin-cushioned on the managerial agenda, which, as Seymour and Rooke argued 'does not require researchers to question their own position. Instead, rationalists put their faith in the use of particular methodological routines to guarantee their impartiality. The researcher's values are regarded as either irrelevant or self-evidently correct'.[153] Furthermore, the egalitarian approach that emphasises mechanistic solutions is inappropriately inept to deal with understanding what is ostensibly a social problem in the delivery of construction projects. There is a need for a broader view such that '[...] the objective of practitioners, for example, quality, efficiency, productivity or profits, cannot be taken to be self-evident by the researcher. An essential purpose of research is to establish what participants in the situation under study, managers, engineers or steelfixers, mean by these terms and what values and beliefs underlie such meanings. Researcher may well share some of the understandings of some of the participants, but it is imperative that they suspend their own understandings. Only by doing so can they allow practitioners to speak for themselves'.[154]

Time and again, we have stressed how the construction industry does not operate as a stand-alone entity. Firms operating within the industry invariably interact with the government as major procurer and regulator of the outputs of the industry. Furthermore, firms have to ensure that whatever they produce meets the needs of the communities they are designed to serve. Therefore, any corporate matter is inextricably linked within a complex web of socio-political concerns. Thus, the problematisation of any corporate phenomenon cannot be detached from an analysis of the role governments and communities play in shaping the agenda. Otherwise, solutions found without accounting for socio-political factors will forever remain partial, and fail to sustain wider, longer term interests of society.

The future of professionalism in the construction industry

Perplexed by where the future of built environment professions lie, the CABE and the Royal Institute of British Architects (RIBA) commissioned a book entitled *The professionals' choice: the future of the built environment professions*.[155] Davies and Knell, in concluding for this book, maintained that 'Professions still hold a unique status in our contemporary institutional landscape, but an unwieldy one. They are notionally self-regulating and independent, yet influenced by market demand and government requirements'.[156] Notwithstanding this, the eclectic range of contributions in this edited book generate a debate on the future role of professional, especially in terms of what is meant by providing good for society.

Jobling[157] discusses the growing complexities in the regulatory framework as a result of the continued expansion legislation (especially EU directives) formulated to address issues of professional liability in dealing with issues like health and safety and the environment. This, in turn, would lead to the need for a plethora of professionals working closely together, and a greater emphasis on professional judgement, to solve design problems. For Curry and Howard,[158] the challenge for professionals in the future is their ability to balance the long-term view of contributing to the good of society and the short-term managerial interests of serving the commercial bottom-line. They stressed, however, that professionals should '[...] bound by their training to do good rather than harm, [hold] a wider view than that of their client's self-interest'.[159] Although this will become increasingly challenging in the perpetuation of the culture of financial interests and performance targets, they conceded that regulations (e.g. the Public Interest Disclosure

Act of 1998 in the UK) can be tightened to protect professionals in meeting their ethical obligations.

In the context where such obligations translate to providing a built environment that users actually want and need, Strelitz[160] suggested that the skills of built environment professionals and the rules by which meaningful building identities are constructed need to be broadened. Accordingly, built environment professionals owe it to the general public to facilitate the process in which increasingly enlightened clients and end-users can be effectively and democratically involved in dialogue about design and construction with them. Besides, 'How are people to have their aspirant meanings invested in identities of building and place, and how is the evolution of the built environment to provide relevant cultural and symbolic linkage, if people at large are not referenced in the development process?'.[161] Hughes warned, however, that the danger lay in the movement towards mass customisation and growing automation of the procurement process where negotiation and decision-making can be made simply by the push of a button, as he predicted the disappearance of the thinking, imaginative professional, replaced by 'graduates who were excellent at routine but unable to exercise judgement'.[162] Of course, Hughes reiterated, few true professionals will survive 'with the skills necessary to interpret needs and to craft a design which can be built by the rare specialists who produce hand-made bespoke work [...] and the things that they do are very special indeed'.[163]

So, professionals in the future are likely to face more pressures to meet ethical and professional codes of conduct if the sanctity of self-governance is to be maintained. Accruing the privileges associated with professionalism is contingent on discharging the moral duty of giving back to society. We have seen how, in this respect, the exclusive nature of professional institutions has thus far prohibited the full potential for professionals to contribute to the public good. Professional arrogance has led to the reinforcement of class boundaries and done little to ensure that products and services are delivered to the highest possible quality. Ironically, in an ever-more professionalised context marked by a proliferation of professional institutions, there is a crisis of regression confronted by professionals. Not only are they increasingly mistrusted, but there is evidence that safeguarding the self-interest of narrowly defined professional boundaries has led to the preservation of outmoded practices. More demands will, however, be placed on professionals in the future as societal dynamics shift on two counts. First, the intensification of consumerism, coupled with higher levels of education, will mean that more discerning end-users will force professionals to engage in new ways of working. Second, as legislation becomes more complex and internationalised, professionals will have to be equipped with new skills to survive. Such new skills include more social and cultural understanding to enable professionals of the future to engage effectively with end-users so that they can democratically co-produce the requirements of the built environment in the future. Furthermore, as more pressure on innovating is exerted on professionals, there will be a drive for professionals to become better in their technical knowledge. In turn, professionals of the future would hopefully be able to regain the respect from contemporary society.

We see once again the criticality of social relations in shaping the past, present and future of professionalism. In the next section, we synthesise the strands of governance in the construction management literature within the overarching theme of human relations.

Connecting political, corporate and community governance in construction: the importance of human relations

A common cross-cutting theme made throughout this book is the importance of human relationships in construction. In Chapter 2, we discussed leadership in the context of personal development of our influential figures and we demonstrated how networks of people and places are instrumental in shaping the lives of our participants. We then outlined various strands of sustainable development and maintained that any perspective should ultimately consider the impacts on people's lives and well being. We have also sought in this chapter to illustrate the power of people to make governance structures in political, corporate and community arenas connect and work for the betterment of industry. Furthermore, it was established that institutional structures and the professional landscape that govern construction should always bring benefits of public good to society.

The review of the governance literature as applied to the field of construction management in the preceding sections has pointed to the interdependencies between organisational structures, technology and people, and how government, corporations and communities can jointly mobilise resources to do things. Far less attention has been placed hitherto on how the grassroots level in communities can be mobilised to contribute to the social and economic value of construction. Nonetheless, we have seen a growing emphasis on the importance of the social at the core of theoretical debates. So, centrally based theories of control and the unitarist perspective of political and corporate governance structures have been criticised for being outdated, as they fail to account for how people at the grassroots level can democratically participate in the shaping of the built environment. The rise of the enlightened customer and end-user, and the impetus of social responsibility, imply a need for advocating more partnerships, especially between the societal stakeholders of government, firms and communities. Devolving power down to the communities might seem inevitable given the breakdown of conventional barriers through such trends as globalisation and the digital age. So, national governments no longer retain control because of the phenomenon of globalisation; and goods and services are no longer provided by corporations as single entities, but rather through complex networks of companies for which power and control get distributed.

One cannot stress sufficiently the importance of human relations in construction. As Nicolini commented 'The importance of interpersonal relations, team spirit and collaboration is a recurring theme in construction management. The take up of innovative procurement and business practices such as partnering, lean construction and supply chain management require the adoption of non-confrontational attitudes, a collaborative spirit, and trust that, in turn, highlight the importance of social, human and cultural factors in the management of construction organizations and projects'.[164] In explaining what good 'project chemistry' meant, Nicolini concluded that '"project chemistry" should not be thought of as a characteristic of the people involved as much as a trait of the relationships between people, task and organizational conditions. Thus, "project chemistry" is a process that needs to be nurtured and maintained throughout the project',[165] thereby reinforcing the point made earlier in this chapter that interorganisational and interpersonal relationships matter!

Pryke and Smyth[166] argued that the relationship approach signifies a timely paradigm shift that centres the attention on the people that deliver projects, how they interact with one another, and how people benefit from having construction. Pryke and Smyth and their contributors address 'fashionable research topics of communication with stakeholders, emotional intelligence, team working, client handling, networks of relationships, discourse'.[167] It is the ability of human beings to socially interact with one another that galvanises the delivery of construction projects, and the success of any project is down to how relationships between one human being and another are forged.[168] Yet the comprehension of how interpersonal relationships can be effectively mobilised remains an under-researched aspect in the construction management literature.[169]

That said, many scholars have acknowledged the importance of human relationships in the workings of the construction sector. Pietroforte, for instance, noted, 'The interactive nature of the design and engineering process, the growing number of participants and the interdependence of building systems demand increased informal communication and mutual adjustment as the coordination mechanism. [...] Successful communication cannot be achieved only through IT investments if the proper attitudes towards cooperation and shared goals are not developed at the onset of projects'.[170]

There is still scope for much more understanding beyond technocratic explanations that predominate in the construction management literature, to accentuate the often-informal interactional aspects of human relations in construction. Furthermore, there is a need for explanations to go beyond the analysis of actions at the individual level to see how human relationships across government, corporate and community actors can be better organised for the betterment of the industry and society. For example, much writing about the industry focuses mainly on the working lives of those involved in the sector, and it is surprising how little has been done to see how working lives and personal lives intertwine to influence developments of the construction industry.[171]

Closing thoughts

In the previous chapter, we articulated the concerns of our leading thinkers regarding their agenda for the future, which we framed in the concept of sustainable development. In many ways, this chapter is about explaining the structures that could potentially govern any action aimed at realising the sustainable development agenda. In other words, if Chapter 3 is about what needs to be done, we are more concerned within this chapter on how this can be achieved. Unsurprisingly, our interviewees have placed much confidence on the power of human actions to engender change for a sustainable future. The analysis of the interviews and the subsequent review of the literature, both in the fields of mainstream governance and construction management, reveal that a complex web of interconnections between government, the private sector and the community matter when trying to design an institutionally coordinated response to address the sustainable development challenges of the future.

The phrase 'institutionally coordinated' has been deliberately chosen. By 'institution', we refer not only to public institutions, but also to how individual actions across the spectrum of government, corporate and community actors can be aligned and disintegrated, configured and reconfigured into an institutionalised way of thinking and acting.[172] This chapter has provided both theoretical and practical insights into the critical issues that need to be accounted for in the journey towards an institutionally coordinated response for a sustainable future. Indeed, we have seen, through both the interviews and the literature review, how the interactions between government, private sector and community actors have aligned their agendas over time. Specifically, we have seen how the tightening of fiscal budgets and the trend of globalisation, alongside ever-more complex societal problems, have meant that engagement with the private sector and communities have become an inevitable reality. We have seen how this development also bears ramifications on the way governments and corporations have transformed in the way they think about, and enact on, critical issues.

However, there is nothing prescriptive in the way the tripartite arrangement can be developed. No silver bullet would be adequate to capture the dynamic relationships between the three stakeholder groups. Besides, one needs to be sensitive about the heterogeneity found in different national contexts across the world, and any rationalistic categories aimed at facilitating comparisons would merely result in an artificial explanation of the structures of engagement at a static point in time. Of course, theoretical endeavours have often devoted much energy to seeking prescriptive mechanisms to achieve joined-up governance between the three key stakeholder groups, framed in the wealth of literature on partnerships and collaborations. At the same time, critical scholars have also alluded to the fact that there are disconnections between the theory of integration and the practice of cooperation. Through the analysis of our interviews, and our discursive critique of the literature, we do acknowledge that reality does not always fit neatly into neat, theoretical categories conveniently proposed by researchers. However, the formulation of discrete categories is not a futile exercise in itself, as it assists in the process of making sense of the world. Again, we see much congruence in what our influential figures have to say and the theoretical issues reviewed in this chapter. Such confluence is not accidental. Ultimately, what our interviewees yearn for is a simplified view of joined-up thinking and working, as they constantly search for greater clarity and transparency in the institutional structures across government, corporate and community actors that govern the work of the construction industry. We have illustrated these points in Figure 4.1 below.

So, what is inhibiting such clarity to be afforded? In this chapter, we have revealed a great deal of tensions that our interviewees often grapple with as they enter into conversations about the emergent future. These tensions manifest in a number of ways, which are elaborated as follows:

- *Time*: there is often the tension between short-term gains and maintaining a long-term view. Whereas all our interviewees have mentioned their desire to secure the longevity of the industry and society at large, their survival instincts often steer them towards offering more pragmatic responses towards meeting present-day challenges. The theoretical literature also frames this paradox as an ethical question relating to realising intergenerational benefits in securing a sustainable future;
- *Place*: the trend of globalisation is often conflictual. Our interviewees talk of the benefits and challenges of globalisation, and although they acknowledge how the global context is crucial in shaping the decisions they make, it is imperative that solutions are found to

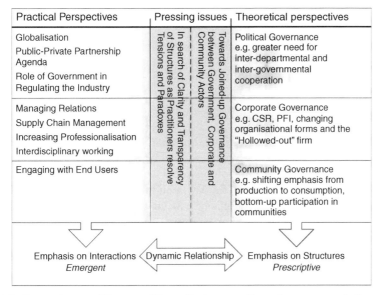

Practical Perspectives	Pressing issues			Theoretical perspectives
Globalisation Public-Private Partnership Agenda Role of Government in Regulating the Industry	In search of Clarity and Transparency Tensions and Paradoxes of Structures as Practitioners resolve	Towards Joined-up Governance between Government, Corporate and Community Actors		Political Governance e.g. greater need for inter-departmental and inter-governmental cooperation
Managing Relations Supply Chain Management Increasing Professionalisation Interdisciplinary working				Corporate Governance e.g. CSR, PFI, changing organisational forms and the "Hollowed-out" firm
Engaging with End Users				Community Governance e.g. shifting emphasis from production to consumption, bottom-up participation in communities

Emphasis on Interactions ⟨ Dynamic Relationship ⟩ Emphasis on Structures

Emergent   *Prescriptive*

Figure 4.1 Comparing practitioner and theoretical perspectives of governance in the context of the construction industry.

satisfy localised requirements. The tensions of thinking-global-acting-local have also been explained in the theoretical literature;

- *Authority*: there is also mention of the paradox of control. Many of our leading figures recognise the inadequacies of top-down authority where the political and professional elite dictate the manner in which the built environment is shaped for the well being of society. However, whereas there is acknowledgement of the importance of engaging with end-users in designing and constructing the built environment, concerns are also raised as to whether bottom-up approaches are necessarily better given the difficulties of gaining consensus and the level of expert knowledge possessed by end-users. The theoretical interpretation as to how the power relations should be configured in this respect remains divided;

- *Representation*: fundamentally, there is the tension between individualism and collectivism when discussing the aspirations of joined-up governance. Although the espoused benefits of collaboration have been acknowledged, there are times when our interviewees consider 'disintegration' to be useful as well. So, for example, when discussing the heterogeneity of the make-up of the sector, a number of participants have noted the clear distinction between large and small firms operating in the industry, and that there should be more recognition that a 'one-size-fits-all' approach is inadequate. Furthermore, many of our influential figures often represent a diverse range of interests within the industry, which can surely be a source of confusion as well. The theoretical literature has also alluded to the problems of seeking appropriate representation, especially when the rules of engagement tend to be framed by an elite few.

It would appear that in organising an institutionally coordinated response, stakeholders within the tripartite arrangement across government, corporations and communities constantly have to grasp the nettle of these contradictions. Arguably, practices by which these ambiguities are resolved would be divergent in reality, which must serve to prohibit any meaningful prescription of what institutional structures should govern the workings of the industry.

It must be added, however, that these tensions are necessarily binary, as thinking in the binary can be useful in decision-making as binary categories enable a quick understanding of the critical issues that matter. Yet, it is uncertain how decision-making can be done precisely and effectively; for all intents and purposes, it is difficult to define what effectiveness means and the framing of this could be divergent over time anyway! What is certain, however, is that the order of the day seems to be explained by the need for joined-up thinking and acting. The future is likely to be inter-everything, from interdisciplinary learning to interprofessional working to intersecting across levels of governance to emphasising social interactions. Again, this is much easier said than done, and there is more scope for research into how notions of power, control and authority are shaped and transformed by such joining-up. These issues rarely get scrutinised in the construction management literature.

Institutions remain in constant flux. They disintegrate and integrate over time, as a result of a rise and fall of new and old actors involved in construction. We have seen how the emphasis on the production function has been replaced by a new light shining on the nature of consumerism. We have discussed how this has meant that end-users have become more empowered to democratically participate in shaping of the outputs produced by the construction industry. We have also discussed how depoliticisation of political governance has led to greater involvement of the private sector, and how the emphasis on the economic imperative is a result of the promulgation of managerial functionalism. We have explained how the inadequacies of government regulation of labour markets has led to a transformation of the modus operandi of the industry, one that is characterised by intense subcontracting and perpetuation of flexible labour markets.

It is, however, anyone's guess as to whether such trends will remain in the future. Perhaps what can be done to construct the future is for practitioners to aspire towards joined-up thinking and working across government, corporate and community actors. As the stakeholders muddle through various arrangements of the tripartite relationship, emergent lessons can be captured to help explain what works effectively and why. Research efforts would be far more fruitful if attention was focused on how the tensions discussed above are conceptualised by those involved, how these contradictions are resolved and how society at large seek to address balance in shaping the future.

Part 3

Towards an afterthought

Chapter 5

The last word: synthesising lessons learnt from the journeys...

'*Work without Hope draws nectar in a sieve,*
And Hope without an object cannot live.'

Samuel Taylor Coleridge, 1772–1834

Chapter summary

In this chapter, we conclude with the key lessons learnt by pulling together the various strands discussed in the preceding chapters. At its core, the concept of emergence is a critical finding. Whether we are discussing the concept of leadership, or framing the sustainable development agenda, or analysing governance structures, it has been noted that deterministic categories of understanding formulated through rational, positivistic approaches are less relevant than examining the dynamics of just how such categories emerge. In a sense, this book has demonstrated that boundaries are always in constant flux. The changing forms of relationships between the constituent parts of construction that we have explained hitherto inevitably changes traditional notions of boundaries. At the same time, boundaries are increasingly being dismantled; this includes the breaking down of geographical boundaries as a result of globalisation, and the convergence of disciplines in the intensification of interprofessional working, and the unifying of academic and industrial discourse in the era of co-production of knowledge.

The discussions presented in the book also show how our interviewees cope with understanding the malleability of how human interactions in the industry help shape the structures that govern its operations, thereby illustrating the interplay between governance structures (Chapter 4) and human agency (Chapter 3). A number of tensions and paradoxes will also be reiterated in this chapter, including how 'futures thinking' resolves the tensions between short-term imperatives and maintaining the long-term view, how bottom-up power helps shape the representational elite's view about the future, how our interviewees resolve thinking about the divisions between individualism and collectivism, and how there is always a need to think-global-act-local. In order to consider future scenarios effectively, this book has presented the case that managers need to engage with

Constructing Futures: Industry Leaders and Futures Thinking in Construction   Paul Chan and Rachel Cooper
© 2011 Paul Chan and Rachel Cooper

the grassroots level, and understand more deeply what people do within construction for the benefit of industry and society.

The key issues discussed in this chapter are as follows:

- 'Futures thinking', in theory and practice, is really 'emergent thinking'. Future scenarios planning in the construction industry is more evolutionary than revolutionary, as policy-makers attempt to frame options for the future in order to cope with the messy realities of today;
- We are living in an increasingly boundary-less world. Although the removal of boundaries can be seen across all walks of political, corporate and community life, boundaries can also be extremely helpful in enabling people to make sense of the world;
- Understanding how people deal with tensions, ambiguities and paradoxes arising from societal change is a critical step in understanding how the future is constructed.

Introduction

We started with a desire to understand the value systems that drive influential people working in the construction industry, and explain how their underlying motivations help contribute to the way they think about the future. We traced this through the personal and career histories of 15 leading figures in the UK construction industry, by establishing how critical people, places and events have influenced the way our participants made sense of the construction industry. We then examined their views on key issues relating to creation of a sustainable future for the construction industry, and their thoughts on how governance structures can be better understood to meet the sustainable development agenda.

In this chapter, we synthesise key conclusions arising from what is an eclectic range of issues discussed both by our interviewees and in the theoretical literature. We first recap the critical points summarised at the end of each preceding chapter. Thereafter, we discuss the main conclusions, which focus on three core messages. First, it is important to remember that 'futures thinking' is an emergent process, and that scenarios derived from the production of foresight studies do not represent objectivised visions of the future to be realised as they are often depicted. Rather, these scenarios are a means for those involved in their formulation to make sense of the tensions and ambiguities of socio-technical change. Second, such societal change invariably brings about disruptions to the established order of things. Clear, traditional boundaries that once helped people cope with organising messy realities of the world increasingly become nebulous. Contemporarily, the disordering of boundaries can be seen through trends of globalisation, blurring of distinction between public and private space, the dismantling of the old order of the corporate firm, and demands for intersections between professional disciplines. Third, such vagaries create tensions and ambiguities that need to be resolved, including trade-offs between economic, environmental and social concerns in sustainable development, and the conflicts that practitioners get embroiled in as they work towards developing a sustainable future for the industry.

Recap on previous chapters

We summarise the salient points learnt in Chapters 2, 3 and 4. We begin by re-examining leadership as an emergent process, before emphasising the missing social link of sustainable development, and discussing the aspirations of joined-up governance.

Leadership as an emergent process: moving away from individualistic notions of leadership in construction

'A young man was walking home one evening after a visit to the local pub. As he came to a subway, he noticed an older man crawling frantically on the ground under a lamppost. The younger man approached the older man and asked if he needed any help. "I've lost my contact lenses", the older man replied. "Where have you misplaced them?" the younger man asked. "Over there!" the older man said, whilst pointing fifty yards away in the dark'.[1]

Like the older man, the definition and understanding of leadership have frequently focused on where the lamppost is. Scholars who have attempted to expose leadership traits and behaviours have falsely illuminated individuals who are presupposed to be leaders. Wood, who suggested that there is a fallacy of misplaced leadership, argued that 'a mistake of management studies is exactly this logic of leadership as if it were "synonymous" with an immediate individuality'. He added, '"successful leaders" are not simple, locatable social actors, nor are they the completion of an operation of individuation. An apparent individuality is construed as a selective abstraction from the vast field of experience. This selective process is prevalent because leaders tend to immerse themselves in a misleading Western "substance metaphysic". They have done this by having certain ascendant characteristics ascribed: I am a visionary, I communicate well, I encourage participation, I build teams, I am clear what needs to be achieved, and so on'.[2] Wood (2005) supported the notion of leadership as an emergent process where the shaping and continuous learning and development of leaders, situated within a community of practice, was more important than individuals.

In many respects, we started this research by pursuing the case of misplaced leadership as our selection of interviewees was based mainly on individuals who held authoritative, influential positions in the UK construction industry. However, whereas many leadership scholars have often glorified leaders on a higher pedestal, we attempted to emphasise more emergent aspects of the dynamics of leadership as applied in the construction industry. We illustrated how our interviewees have worked, even persevered, to the top. And many were keen to dismiss the idea that they deserved the title of a construction leader. As Guy Hazelhurst reinforced, all he wanted to do was to make 'an impact'. We have also shown a more collective view of leadership that transcends the individualism that many scholars accord to the subjects. We saw how people, places and events helped shape the thoughts and practices of our influential interviewees in the construction industry. It is the situated practice of engaging within their communities that offered our interviewees the range of experiences that continually and dynamically define how they respond to the challenges of the day. We emphasised the role of learning in the enactment of leadership and the significant role social dynamics play in helping us understand how leadership happens beyond individual constructs. Arguably, life history methodology enabled us to drill deeper into the personal value systems of our interviewees so that we can articulate how these might shape the future of the industry they clearly influence.

The missing social link of sustainable development

Conversations about the future certainly evoked thinking about sustainability matters, which were reported in Chapter 3. It was interesting to observe a great deal of congruence between our interviewees' recognition of what sustainability means and the theoretical explanations of sustainable development. There is indeed the understanding that to secure a sustainable future for human societies requires the balancing of economic, social and environmental concerns. Moreover, new skills are demanded to help professionals and new entrants navigate through the complex minefield of interrelated concepts of sustainable development. Education and research play a critical part in this, as knowledge about sustainable development is still developing.

It was also useful to find that our interviewees all reiterated the importance of humans when discussing sustainability. Still, the social dimension remains elusive, given the dominance of the economic perspective that prevails in debates about sustainable development. Paradoxically, framing sustainable development in economic terms was found necessary to arm politicians, business and community leaders with the requisite language to discuss impacts of, and interventions on, meeting the sustainable development agenda; yet there was acknowledgement that the emphasis on monetary valuation prevents genuine progress made to secure real benefits of a sustainable future. The economic perspective results in the obsession with a top-down measurement approach, which places less emphasis on the qualitative nature of the future well being of people. Furthermore, the obsession with defining and measuring the sustainable development agenda detracts from material concerns on how policy formulation and implementation can be done effectively.

Sustainable development is a complex concept encompassing a number of interconnected facets, which we explained as a complex web of man-made capital (economic), social capital (social), natural capital (environmental) and human capital (skills) perspectives. Understanding trade-offs in decision-making is, therefore, critical, especially where there is a fragmented landscape of stakeholders involved. There is a need to consider the socio-political and economic structures of decision-making, and opportunities for joined-up thinking and action need to be explored. The social dimension needs to be brought more to the fore. Sociological and psychological disciplinary knowledge can be mobilised to understand the nature of human agency and behavioural change in delivering the sustainable development agenda. To do this, there is a need to consider how an institutionally coordinated response can be harnessed by integrating government, corporate and community actors.

The aspirations of joined-up governance

In Chapter 4, our interviewees commented on a number of significant societal trends, including the force of globalisation that has both positive and negative ramifications for how the industry operates in the future, as well as the shedding of public responsibility to the private sector for infrastructure development, and the changing role of education, research and professionalism and how these contribute to the benefit of the industry and society at large. What came out from the analysis of our interviewees' perspectives is the blurring of boundaries and the dismantling of the established order of doing things. These imply a need for better understanding and coordination of the social interactions between the key stakeholders of government, private sector and community actors. Indeed, an institutionally coordinated response to delivering a sustainable future demands joined-up working across political, corporate and community governance structures. It is imperative that better ways of arranging the relationships between the key stakeholders can be designed to deliver well being across the various strands of sustainable development discussed in Chapter 3.

However, institutional arrangements have been found to be somewhat fluid, and are contingent on the context in which the arrangements are mobilised. Our interviewees suggested that a 'one-sizes-fit-all' model is likely to fail, and that one needs to seek localised solutions to meet localised needs, but within an increasingly globalised context. Institutional arrangements also imbue a sense of power relations, and it is critical to note how the power has shifted more to consumers and end-users of the goods and services generated by the industry. Indeed, institutions disintegrate and integrate over time, as a result of a rise and fall of new and old actors, respectively, involved in the production and use of the built environment.

Calls for joined-up governance have been raised, both by our interviewees and in the literature on governance. It is, however, argued that joined-up governance is merely an aspirational concept for stakeholders to work towards. After all, there are a number of tensions and paradoxes that have to be constantly resolved. These include balancing short-term demands of the marketplace with the need to maintain a long-term view, the tensions between globalisation and localisation, and the ambiguities of power associated with devolving policy choices to the grassroots. These contradictions will be further discussed in greater depth in the next section. However, what is clear is that how stakeholders resolve these tensions is likely to be divergent and contextual. Therefore, instead of prescribing how the actors across government, the private sector and community can be coordinated to effect change for a sustainable future, perhaps a plausible recommendation is for researchers to contribute to a better comprehension of how governance structures, and related notions of power, control and authority, evolve over time.

Key conclusions

The conclusions of this book centre on three fundamental themes of emergence, boundaries and contradictions. We first assert that 'futures thinking' in construction really is an extension of present-day thinking. Rather than treating this as revolutionary, 'futures thinking' is about stakeholders associated with the industry trying to make sense of the changing world around them. Second, we explore the notion of boundaries, and discuss how boundaries are consistently being disrupted in the age of hybrids. Finally, we focus our attention on what this means for the professional who has to deal with ever-more complex contradictions.

Futures thinking as emergent thinking

When we reviewed the various foresight studies at the beginning of this book, we asserted that the future scenarios reported in these studies tended to paint an objectivised view of the future. It was suggested that the committees – represented by an elite few in academia, government and corporate policy-makers – create knowledge about these future scenarios by formulating rationalistic and deterministic categories about what future trends might look like. We maintained that such offerings are limited in terms of engendering real action for change, as categories themselves mean very little without the context by which such futures can be implemented and enacted. In fact, we would argue that the identification of future scenarios is really more about providing policy-makers an opportunity to make sense of options given the uncertainties of change that can often be confounding. Taking this view, therefore, 'futures thinking' in the construction industry is less about revolutionary stuff than evolutionary thinking. Thus, it is not surprising to find so much convergence in the elicitation of critical themes across several foresight reports and the views of the future that our interviewees held. Future-gazing in construction is never fictional, but more about seeking often pragmatic consensus in a turbulent world.

So, if foresight studies are not intended to lead to any real consequence, at least immediately, then why bother? One compelling reason is that foresight studies are a mechanism by which policy-makers can initiate change in the construction industry. In Chapter 2, we certainly highlighted the fact that many of our influential figures are mavericks and rebels who are often dissatisfied with the status quo. As such, engaging in 'futures thinking' allows them to become change-makers in the industry. A more conservative explanation is that foresight studies offer a useful commentary on societal change. By organising messy realities into contemporary future trends, these studies provide a frame of reference through which people (well, the policy-makers sitting round the table at least) can engage in a meaningful discussion. As we argued earlier, this aids the sense-making process of what can be rather complex socio-technical change. Besides, future scenarios promise a sense of renewal, a fresher perspective of agenda passed. We observed in Chapter 1 that the broad agenda framed in many of the post-war inquiries into the construction industry are only subtly different from the broad agenda of today, and, indeed, tomorrow. What has changed are specific emphases and the introduction of new frames of reference with which societies act. It is, therefore, vital that research is undertaken to continually capture and critique the fundamental lessons learnt as 'futures thinking' evolves over time.

Disrupting boundaries: the age of hybrids

Change can be terribly disruptive! Throughout this book, we have witnessed funda-mental societal change in a number of areas that have or would have an impact on the construction industry. We have seen, for instance, discussions on globalisation and how this can, on the one hand, enrich both economic and labour markets, but also bring in a new world order in the shape of emerging powers like China that could perhaps disrupt the localised view of construction. We have also seen how boundaries between government, the private sector and communities are starting to be dismantled in the name of joined-up governance. Although this is thought to bring out benefits of empowerment, corporate social responsibility and a more efficient way to manage fiscal budgets, we are still a long way away from finding optimal ways of managing this tripartite relationship as we move towards an institutionally coordinated response for a sustainable future. Indeed, joined-up governance is all but an aspiration. Research efforts into understanding how we can collectively get there are probably more rewarding than reinforcing the partial truths of rhetoric and prescription that predominate much of the management literature.

We have also seen more calls for interdisciplinary, interprofessional and intersectoral working. Whereas there are potential benefits of innovation to be reaped by intersecting disciplinary and professional knowledge, there are still tensions created as a result of protectionist behaviour exhibited by those who represent disciplinary, professional and sectoral interests. Furthermore, divisions between academia and industry appear to be overcome, at least at the strategic level of the industry. There is definitely a lot of confluence between what our interviewees had to say and what the theoretical literature has to say.[3] Still, more needs to be done to figure out how academia and industry can better work together to co-produce knowledge of the future. So, in the age of hybrids, established ways of doing things get replaced by the new order of crossing boundaries.

The breaking down of barriers, however, brings about contradictions. Although disrupting boundaries offers the promise of finding better, more efficient ways of doing things, old boundaries themselves can also provide a sense of protection for those who either resist or are uncertain about change. We see this with the example of technological advancement. There will always be some who will be forever optimistic about technological progress, and there will always be others who will be cautious, even anxious, about the social implications of such innovations. Moreover, stripping down boundaries does not always offer a universal panacea to the challenges confronting the industry. Boundaries do exist for a purpose, as Chris Luebkeman reiterated, 'Good fences make good neighbours'. The removal of boundaries does imply a greater need for coordination, especially in a world characterised by collaborations. How such collaborations can be effectively coordinated remains a significant area of inquiry. Arguably, research can usefully contribute by recording how change takes place, how boundaries between the old and the new alter over time, and analysing the implications on the construction industry and society.

Tensions, ambiguities and paradoxes

We have equated 'futures thinking' with 'emergent thinking' and we have noted the fluidity by which boundaries chop and change over time. Such vagaries consequently result in a number of tensions, ambiguities and paradoxes with which practitioners in the industry constantly have to grapple. For instance, we have discussed the tensions faced when trading off economic, social and environmental constraints when addressing the sustainable development agenda. More research needs to be undertaken to explain how decisions are made about trade-offs, and how the inherent tensions found in the triple-bottom-line agenda are resolved by those working at the coalface of the industry.

Another critical tension relates to time. In Chapter 1, we observed that the foresight reports reviewed have opted for a 15–20-year time-horizon when discussing future scenarios. Reflecting on the interviews, there is some ambiguity in terms of how our interviewees have framed the notion of time. Suggestions of a 5–10-year horizon were made by a couple of our leading figures. This, in itself, is rather interesting as it further reinforces the argument that 'futures thinking' is not some fictional idea in the distant future, but one that is rooted in a shorter term view connected with a urgent sense of pragmatism. It is also worth noting that foresight reports, undertaken collectively in committees, have tended to take a longer term view as compared to our leading figures who were individually interviewed. Perhaps there is much to be said about the consequences of collective decision-making here.

In the conversations, our interviewees often struggled with the tensions of balancing short-term pressures of the marketplace and the need to maintain a longer term view. It would appear that they considered the latter as an ideal to aspire towards, while recognising the primal need to survive in the present context. This perhaps explains why 'futures thinking' in the construction industry is more incremental than radical. As Machiavelli once remarked, change is so difficult because those who stand to lose are most certain of their loss, whereas those who will benefit are uncertain of their gain. Thus, there is a tendency for people to stick to a short-term view, even when thinking about a sustainable future. Is this necessarily conservatism or is this inevitable pragmatism that provides a sense of balance? Indeed, more needs to be done to better understand how the paradox of time is being resolved by practitioners.

Then there are the ambiguities of authority, especially given the changing nature of governance structures discussed in Chapter 4. Here, the tensions operate across a number of levels. There is, for instance, the paradox of holding a globalised view, while maintaining localised action. So, the desire for freedom of capital, and even labour, mobility can also be met with resistance and protectionist behaviour that is not atypical in the current global financial crisis. We have seen this ingrained in the passion for localism maintained by our interviewees as they provided a balanced response to the challenges of globalisation.

The paradox of authority is also demonstrated in the shifting power relations observed when we discussed the evolution of governance structures. For example, the depoliticisation process meant that devolution of power from public administrators to the private sector in providing for infrastructure development is met by the retention of greater control by the state. The same can be said for the tensions arising from the devolution of power from central governments to the regional and local authorities. Furthermore, we have seen contradictions raised when our interviewees talked about the rising significance of end-users in shaping the built environment. In shifting the emphasis from political governance to corporate governance to community governance, there is the tension of defining where power, control and authority lie.

Is the approach of top-down decision-making by policy-makers and professionals really irrelevant in the age of consumer sovereignty? Are consumers really empowered, and indeed enlightened enough to make a more bottom-up approach effective? Are all end-users consulted anyway? Indeed, this last question brings yet another tension of representation and the contradictions between individualism and collectivism. How do we ensure that individuals are appropriately and adequately represented in the collective? And what is the collective anyway? Examining how practitioners go about resolving these tensions of authority will certainly make a fertile ground for further inquiry.

Knowledge gaps to frame the research and practice agenda of the future

To conclude, therefore, there are a number of questions that remain unresolved, which could form agendas for research and practice in the future. These include:

- How can researchers and practitioners work together to capture and critique the process in which future trends of the industry emerges? What are the methodological issues that need to be considered to do this? And how will the findings of such research be utilised?
- In the age of hybrids where interorganisational relations matter more than ever, how do researchers and practitioners conceptualise the notion of boundaries? How can we track the constant disordering of boundaries effectively? And how can we translate this to, and for, practice?
- How do we configure and re-configure power relations in the industry? And how do we recognise the rise and fall of new and old societal actors that help shape the construction industry?
- In examining tensions, ambiguities and paradoxes, we have mainly illustrated these in binary terms, i.e. short term versus long term, global versus local, top-down versus bottom-up, and individualism versus collectivism. Binary thinking indeed simplifies the understanding for both researchers and practitioners. Furthermore, binary thinking probably expedites decision-making. But are there issues missing by merely concentrating on the binary?
- When investigating decision-making, whether at the political, corporate or community levels, how can researchers and practitioners know of the unintended consequences?
- We have concluded that 'futures thinking' is more incremental than radical in the construction industry. How (and why) has this changed over time? And how is it similar or different to other industrial sectors?

Epilogue

When we first embarked on this book project, like the art of crystal-ball gazing, we did not know what to expect. All we wanted to do initially was to speak to a number of influential people in the UK construction industry to learn from their personal journeys and thoughts about the future. The anticipation was that rewards could be reaped in shaping our thinking about the future, as well as framing research agendas that matter! The conversations we had were fun, and, at times, enlightening. We certainly felt a sense of growing up and maturing through the interactions and analysis of the data generated with our interviewees. It would be tempting to suggest that we deliberately chose the title *Constructing futures* because we knew from the outset that 'futures thinking' was really emergent, and that the critical focus ought to be on the process of *constructing* the future, as opposed to concentrating the attention on *constructed* futures. However, we are not going to deny the process we went through; the collection of the data, and indeed its sense-making were in fact rather unwieldy. In many respects, the process of getting to *Constructing futures* is similar to a typical construction project, where participants cope with both chaotic and orderly aspects of life to generate something somewhat tangible. We hope you have enjoyed reading many of the direct quotes that we re-presented from our interviews. We have attempted to offer a scholarly approach to the comprehension of multiple aspects relating to the future. Finally, this book is intended to initiate a first step in providing a more scholarly reflection to the journey of constructing a future for the construction industry.

Appendix

Brief biographies of influential figures interviewed

Alan Ritchie

Alan Ritchie is General Secretary of the Union of Construction, Allied Trades and Technicians (UCATT). A joiner by trade, his moment of epiphany in the trade union movement came at the age of 18, when he was heavily involved in the bitter dispute in the early 1970s when working in the shipyards on Glasgow's River Clyde. Alan has an impressive track record in the British trade union movement, having been Chairman of the Scottish Trade Union Congress (TUC) Youth Advisory Committee in his early years to his rapid rise to the top in October 2004. Alan Ritchie is extremely passionate about sustaining the skills capacity of the construction industry and is an avid campaigner for a return to direct employment in construction.

Bob White

Bob White is a founder member and Chairman of Mace. In the late 1960s, Bob graduated from the University of Sheffield with a degree in architecture. His initiation into construction site management came through his experience with Nottinghamshire County Council, when he specialised in applied research into building components and systems. Bob was instrumental in delivering many of the innovative systems used on the Broadgate project in the City of London in the late 1980s. Since the early days of Mace, he has been largely responsible for strategic change within the company, including the early broadening of the service offer into project management, and the development of the organisational framework. He is also passionate about human resource development in construction, having championed the development of Mace's *People*. Bob was former Chair of nCRISP, and Constructing Excellence, the best practice club for the UK construction industry.

Constructing Futures: Industry Leaders and Futures Thinking in Construction   Paul Chan and Rachel Cooper
© 2011 Paul Chan and Rachel Cooper

Chris Blythe

Chris Blythe was appointed Chief Executive Officer of the Chartered Institute of Building (CIOB) in January 2000. He is a firm champion of encouraging non-cognates who are working in the industry to gain professional recognition with the CIOB. A non-cognate himself, Chris is a Fellow of the Chartered Institute of Management Accountants. He has 15 years' experience working in accountancy and finance, working for companies such as Dunlop, Birmid Qualcast, Mitel, W Canning, Corgi Toys and GKN. He is passionate about vocational education and training matters.

Chris Luebkeman

Chris Luebkeman runs the Global Foresight and Innovation initiative at Arup, a global design and engineering firm and leading creative force behind many of the world's most innovative projects and structures. He works extensively with some of the world's largest companies to develop what he calls 'plausible futures', to better understand the opportunities that change is creating for them in the built environment. Educated as a geologist, structural engineer, and architect, Chris has a background in both design and research. He has worked as a faculty member at the Swiss Federal Institute of Technology, University of Oregon, the Chinese University of Hong Kong and MIT. Chris has advised the UK Government's Engineering and Physical Sciences Research Council on strategic matters relating to the Built Environment. Chris's biggest wish is that everyone on the planet would not only know, but also be able to understand, the impact that his/her everyday life has on the planet.

George Ferguson

George Ferguson co-founded Ferguson Mann Architects in 1978 and founded the national UK wide network of practices in 1986. He has a wide variety of experience in architectural, master planning and regeneration projects, commencing with regeneration and historic building work that formed the foundation of the practice. This includes his mould-breaking Tobacco Factory mixed-use project and many other award-winning schemes involving new and old buildings. George is a lateral thinker and has been a prime mover for change in attitude to planning and redevelopment in Bristol and beyond. This led him to being elected RIBA President (2003–2005) when he was noted for championing the causes of education, the environment and good urbanism. A co-founder of the Academy of Urbanism, he writes, broadcasts and lectures extensively on environmental politics, planning and architectural matters. He is, among many other things, a Trustee of the London-based international think tank *Demos.*

Guy Hazelhurst

Guy Hazelhurst is Director of Skills Strategy for London 2012. He has had a very varied career pathway. Starting out work as a site agent on building sites after leaving school at 17, he has worked for companies including Taylor Woodrow, Pochins and Fairclough (now Amec). He later returned to undertake a degree qualification at what is now University of West of England in Bristol, UK. He then worked as an academic researcher before becoming a consultant with the Davis Langdon Consultancy. Guy was instrumental in setting up the ConstructionSkills Network, which presents periodic labour market intelligence for the various regions in the UK construction industry.

Jon Rouse

Jon Rouse was appointed Chief Executive Officer of the London Borough of Croydon in 2007, the ninth biggest unitary authority in England. He was formerly Chief Executive of social homes agency, the Housing Corporation, and he also set up and led the advisory body, the Commission for Architecture and the Built Environment (CABE). Rouse worked as secretary to the Urban Task Force in 1998, where he contributed to the production of the report *Towards an urban renaissance.*

Kenneth Yeang

Kenneth Yeang is Director of Llewellyn Davies Yeang, an architectural practice based in London with offices in Malaysia, the USA and Spain. Kenneth undertook Ph.D. research at Cambridge University into the incorporation of ecological considerations in the design and planning of the built environment. His latest book *Eco Masterplanning* offers an insight into his state-of-the-art approach to master planning based on environmental principles. Ken is an inventive and prolific architect, who has focused much of his professional life on designing the bioclimatic skyscraper, a tall building, the architecture of which derives from a systematic understanding of the role climate can play in finding forms and technologies that are energy-efficient, integrated in the city grid and that enhance the quality of life of inhabitants in the tropical city. He has designed skyscrapers in London, Singapore, Kuwait, Canada, China, Turkey, Kazakhstan and Uzbekistan.

Kevin McCloud

Kevin McCloud is a designer, presenter and author. With a family background in engineering, a degree in the history of art and architecture and experience as a theatre designer, he brings many different perspectives to the *Grand Designs* series. Kevin has written books on colour, home decorating and lighting. He believes that architecture and the design of the built environment need to respond to the people who use them. He believes that the best design – whether traditional or radically modern – relates to context: landscape, place and neighbouring buildings. He is also keen on the environmental agenda.

Nick Raynsford

Nick Raynsford has been the Member of Parliament (MP) for Greenwich and Woolwich since 1997 (and for Greenwich from 1992). He joined the Government in 1997 and held responsibility for housing, planning and construction, as well as being Minister for London. He was Minister for Local and Regional Government in the Office of the Deputy Prime Minister from 2001 to 2005. In 2001, Nick was made a privy councillor in the New Year's Honours. He left Government in 2005. Nick is currently President of Constructionarium Limited, which was set up to provide university students studying construction and civil engineering degree programmes to gain practical skills through a residential field visit where students construct scaled-down versions of real-life iconic buildings.

Sandi Rhys Jones

Sandi Rhys Jones has 35 years experience in strategic marketing, management and communications, working for contractors, consultants, suppliers, representative organisations, and national and local government. A passionate advocate for the construction and engineering industry, and, in particular, the role of women, Sandi chaired the government/industry working group on equal opportunities in the UK construction industry following the Latham report in 1994, which highlighted the lack of women in construction. She is currently Non-Executive Director of Engineering UK and Deputy Chair of the UK Resource Centre for Women in Science, Engineering and Technology. Between 2002 and 2009 she was a non-executive director of the Simons Group, a construction, design and property company. She had a particular remit to increase and support women in the company, establishing a number of programmes including an innovative mentoring scheme with clients. She has an M.Sc. degree in Construction Law & Arbitration from King's College London. She is a Fellow of the Chartered Institute of Building, an Associate of the Chartered Institute of Arbitrators and a trained mediator. In 1998, Sandi was awarded the OBE for promoting women in construction, and was awarded an Honorary Doctorate by Sheffield Hallam University in 2005.

Sir Michael Latham

Sir Michael Latham read history at Cambridge University and obtained a Certificate in Education from Oxford University in 1965. From 1965, he served in the Conservative Party's Research Department, before moving into the National Federation of Building Trades Employers in 1967 to begin an engagement with the construction industry that led to chairing the inquiry that led to the publication of the well-quoted Latham Report *Constructing the team* in 1994. Sir Michael continues to play a leading role in the industry as Chairman of Construction Skills, the Sector Skills Council for Construction and as Deputy Chairman of Willmott Dixon Limited. Sir Michael has received many honours, with honorary degrees including from Leicester, Nottingham Trent, Birmingham,

Loughborough and Northumbria universities. He is an honorary fellow of the prestigious Royal Academy of Engineering, the Royal Institute of British Architects, the Institution of Civil Engineers, the Chartered Institute of Building, the Chartered Institute of Purchasing & Supply, the Institute of Building Control, the Architects & Surveyors Institute, the Royal Incorporation of Architects in Scotland and the Landscape Institute and is also an honorary member of the Royal Institution of Chartered Surveyors. He was knighted in 1993.

Stef Stefanou

Stef Stefanou is Chairman of John Doyle PLC, a specialist contracting group with headquarters in Welwyn Garden City, UK. Stef was born in Egypt and is of Greek origin. He has extensive experience in the industry, having worked with Peter Lind, Fairweather and Holst (now Vinci) before joining Doyle PLC in 1972. Stef maintains a high profile within the industry via his involvement in a number of industry-related bodies, including Construct (the association for concrete frame contractors), as Board Member of SpeCC, the registration scheme, Chairman of Constructionarium and Member of the Construction Skills Network Observatory for the Yorkshire and Humber areas. In 2004, he was made a Fellow of the Institution of Civil Engineers. In 2005, the Dundee International Congress awarded him the Concrete Sector Gold Medal. In 2007, he was given the OBE for his services to the Construction Industry and also an honorary Doctorate by the University of Westminster.

Tom Bloxham

Tom Bloxham is Chairman, joint founder and major shareholder of award-winning regeneration company Urban Splash, which has won 272 awards to date for architecture, regeneration, design and business success. In 2008, he was elected as Chancellor of the University of Manchester and is a trustee of the Manchester United Foundation Charity. In 2009, he was appointed by the Prime Minister as a Trustee of the Tate. Tom was awarded an MBE in 1999 for his services to Architecture and Urban Regeneration. He is an Honorary Fellow of the Royal Institute of British Architects (RIBA), and has received a number of honorary degrees and doctorates, including Doctor of Design from University of Bristol and an honorary degree from University of Manchester in 2007, and honorary degree of Doctor of Design from Oxford Brookes University in 2004. Tom started out selling fire extinguishers door to door, then, while at Manchester University studying politics and history, started selling old records and posters from market stalls. He established and subsequently sold the Baa Bar chain, as well as a local radio station. In 1993, he founded Urban Splash with Jonathan Falkingham to redevelop an unloved building in Liverpool, using great architecture into the successful Concert Square mixed-use scheme. Since then, Urban Splash has undertaken more than 60 schemes, creating thousands of new homes and jobs and investing hundreds of millions of pounds into successful regeneration projects across the country including in Manchester, Liverpool, Birmingham, Leeds, Bradford, Sheffield, Bristol, Plymouth and Morecambe.

Wayne Hemingway

Wayne Hemingway is co-founder of fashion label Red or Dead and HemingwayDesign. After 21 consecutive seasons on the catwalk at London Fashion Week, he and his wife sold Red or Dead in a multi-million cash sale in 1999. He then set up HemingwayDesign, joining forces with building firm Wimpey to work on various housing projects specialising in affordable and social design. Wayne is the Chairman of Building for Life, CABE (Commission for Architecture and The Built Environment) a funded organisation that promotes excellence in the quality of design of new housing. He is a Professor at the School of the Built Environment of Northumbria University, and a writer for architectural and housing publications, as well as a judge of international design competitions including the regeneration of Byker in Newcastle and Salford in Greater Manchester. Wayne started his career at London's fashionable Camden Market, selling second-hand clothes. An early boost by an order from the New York department store Macy's, and the company became famous for its footwear, especially its revival of Doc Marten boots. He and his wife's innovative approach saw their designs and stores spread quickly across the UK, then the world, winning them a host of industry accolades along the way. Wayne was awarded an MBE in 2006. Wayne Hemingway passionately believes in the supremacy of design, whether in clothes or in buildings.

Notes and references

Chapter 1

1. Grossman, L. (2004) Forward thinking. *In*: *Time magazine*, 25 October, 46–52.
2. See Murray, M. and Langford, D. (2003) *Construction reports: 1944–98*. Edited text. Oxford: Blackwell Publishing.
3. Edkins, A. (2000: 4) *Building future scenarios*. London: CRISP.
4. Murray, M. and Langford, D. (2003).
5. Fairclough, J. (2002) *Rethinking construction innovation and research*. London: Department of Trade and Industry.
6. Edkins, A. (2000).
7. Broyd, T. (2001) *Constructing the future*. London: CRISP.
8. Gann, D. (2003) *Nanotechnology and implications for products and processes*. London: CRISP.
9. Schwartz, P., Leyden, P. And Hyatt, J. (2000) *The long boom: a vision for the coming age of prosperity*. New York: Harper Collins.
10. Edkins, A. (2003).
11. Simmonds, P. and Clark, J. (1999) *UK construction 2010: future trends and issues briefing paper*. London: CIRIA.
12. Royal Institution of British Architects (2003) *The professionals' choice: the future of the built environment professionals*. Foxell, S. (Ed.), London: RIBA.
13. Construction Industry Institute (1999) *Vision 2020*. Texas: CII.
14. Building futures council (2000) *The future of the design and construction industry (projection to 2015)*. Washington D.C.: CERF.
15. The American Society of Civil Engineers (2007) *The vision for civil engineering in 2025*. Virginia: ASCE.
16. The American Society of Civil Engineers (2009) *Achieving the vision for civil engineering in 2025: a roadmap for the profession*. Virginia: ASCE.
17. Flanagan, R., Jewell, C., Larsson, B. and Sfeir, C. (2000) *Vision 2020*. Gothenberg, Sweden: Chalmers University.
18. Hampson, K. and Brandon, P. (2004) *Construction 2020: a vision for Australia's property and construction industry*. Brisbane, Australia: CRC.

19. Adapted from: Chan, P., Abbott, C., Cooper, R. and Aouad, G. (2005: 716) Building future scenarios: a reflection for the research agenda. *In*: Khosrowshahi, F. (Ed.) *Proceedings of the Twenty-First Annual ARCOM Conference,* 7–9 September 2005, School of Oriental and African Studies (SOAS) University of London, Association of Researchers in Construction Management. 709–719.
20. Harty, C., Goodier, C., Soetanto, R., Austin, S., Dainty, A., and Price, A.D. F. (2007). The futures of construction: a critical review of construction future studies. *Construction Management and Economics,* **25**, 477–493.
21. Chan *et al.* (2005).
22. Cooper, C. (1985) *Change makers: their influence on British business and industry.* London: Harper and Row.
23. Jennings, R., Cox, C. and Cooper, C. L. (1994) *Business elites: the psychology of entrepreneurs and intrapreneurs.* London: Routledge.
24. Silverman, D. (2001) *Interpreting qualitative data: methods for analysing talk, text and interaction.* 2 Ed. Wiltshire: Sage Publications Ltd.
25. The reader is also referred to recent work by Collier (2007, 2009), who illustrated in great detail the life histories of construction 'leaders', both past and present, to demonstrate their contribution to the structure and fabric of society. In doing so, she endeavoured to pay more attention to the stories of the often-neglected 'heroes' in construction, i.e. the builders. However, as far as it is known, this book is the first attempt to explain how the value systems of 'leaders' in construction contribute to the way they think about the future of industry and society. See Collier, C. (2009) *Building visionaries: the unsung heroes.* Englemere: CIOB, and; Collier, C. (2007) *CIOB Construction leaders.* Englemere: CIOB.
26. Musson, G. (2004) Life histories. *In*: Cassell, C. and Symon, G. (Eds.) *Essential guide to qualitative methods in organizational research.* London: Sage Publications. pp. 34–44.

Chapter 2

1. Handy, C. (2001: 17) *The elephant and the flea: looking backwards to the future.* London: Hutchinson.
2. See, for example, Tannenbaum, R. and Schmidt, W. H. (1958) How to choose a leadership pattern. *Harvard Business Review,* **36**(2), 95–101; Blake, R. R. and Mouton, J. S. (1964) *The managerial grid,* Houston TX.: Gulf; and Fiedler, F. E. A. (1967) *Theory of leadership effectiveness.* USA: McGraw-Hill.
3. See Fairholm, M. R. (2004) Different perspectives on the practice of leadership. *Public Administration Review,* **64**(5), 577–590.
4. See, for example, Hammer, M. and Champy, J. (2001) *Reengineering the corporation: a manifesto for business revolution.* London: Nicholas Brealey Publishing.
5. de Vries, K. and Manfred, F. R. (1997) Leadership mystique. *In*: Grint, K. (Ed.) *Leadership: classical, contemporary, and critical approaches.* Oxford: Oxford University Press.
6. Berry, A. J. and Cartwright, S. (2000: 348) Leadership: a critical construction. *Leadership and Organizational Development Journal,* **21**(7), 342–349.

7. Kotter, J. (1990) What leaders really do. *Harvard Business Review*, May–June, **68**(3), 103–111.
8. Munshi, N., Oke, A., Stafylarakis, M., Puranam, P., Towells, S., Möslein, K. and Neely, A. (2005: 6) *Leading for innovation*. AIM Executive briefings, London: Advanced Institute of Management Research (AIM).
9. See Hammer and Champy, op. cit., and; Morrell, M. and Capparell, S. (2001) *Shackleton's way: leadership lessons from the Great Antarctic explorer*. London: Nicholas Brealey Publishing.
10. Kotter (1990).
11. Grint, K. (2005: 1472) Problems, problems, problems: the social construction of 'leadership'. *Human Relations*, **58**(11), 1467–1494.
12. Fairholm (2004: 578).
13. Ibid.
14. Berry and Cartwright, op. cit., p. 344.
15. Ibid.
16. See Munshi *et al.* (2005).
17. See, for example, Stogdill, R. M. (1974) *Handbook of Leadership. A survey of theory and research*, New York: Free Press; Covey, S. (1990) *Principle centred leadership*. New York: Simon and Schuster, and; Yukl, G.A. (2005) *Leadership in organizations*, 6th Ed. London: Prentice Hall.
18. Wright, P. (1996) *Managerial Leadership*, London: Routledge.
19. Fairholm (2004: 579).
20. See Tannenbaum and Schmidt, op. cit.; Blake and Mouton, op. cit., and; Fiedler, op. cit.
21. Grint (2005: 1490).
22. Ibid.
23. Burns, J. M. (1978) *Leadership*. New York: Harper and Row.
24. Strange, J. M. and Mumford, M. D. (2002) The origins of vision: charismatic versus ideological leadership. *Leadership Quarterly*, **13**(4), 343–377.
25. Wright (1996).
26. See, for example, Gronn, P. (2002) Distributed leadership as a unit of analysis. *Leadership Quarterly*, **13**(4), 423–452, and; Spillane, J. (2006) *Distributed leadership*. San Francisco, USA: Jossey-Bass Publishers.
27. Spillane (2006).
28. Munshi *et al.* (2005).
29. Senge, P. M. (1990). *The art and practise of the learning organisation*. London: Century Business.
30. Huff, A. S. and Möslein, K. (2004). An agenda for understanding individual leadership in corporate leadership systems. *In*: Cooper, C. (Ed.) *Leadership and management in the 21st century: business challenges of the future*. Oxford: Oxford University Press. pp. 248–270.
31. Munshi *et al.* (2005: 6).
32. Collinson, D. (2005) Dialectics of leadership. *Human Relations*, **58**(11), 1419–1442.
33. Giddens, A. (1984) *The constitution of society*. Cambridge: Polity.
34. Collinson (2005: 1419).

35. Odusami, K. T., Ivagba, R. R. O. and Omirin, M. M. (2003) The relationship between project leadership, team composition and construction project performance in Nigeria. *International Journal of Project Management*, **21**(7), 519–527.
36. Slevin, D. P. and Pinto, J. K. (1988) Leadership, motivation and the project manager. *In*: Cleland, D. O. and King, W. R. (Eds) *Project management handbook*. New York: Van Nostrand Reinhold. 739–755; see also Lewin, K., Lippitt, R. and White, R. K. (1939) Patterns of aggressive behaviour in experimentally created 'social climates'. *Journal of Social Psychology*, **10**, 271–299, and; White, R. K. and Lippitt, R. (1960) *Autocracy and democracy: an experimental inquiry*. New York: Harper.
37. Nam, C. H. and Tatum, C. B. (1997) Leaders and champions for construction innovation. *Construction Management and Economics*, **15**(3), 259–270.
38. McCabe, S., Rooke, J., Seymour, D. and Brown, P. (1998) Quality managers, authority and leadership. *Construction Management and Economics*, **16**(4), 447–457.
39. Berry and Carwright (2000).
40. Low, S. P. (1995) Lao Tzu's *Tao Te Ching* and its relevance to project leadership in construction. *International Journal of Project Management*, **13**(5), 295–302.
41. Fellows, R., Liu, A. and Fong, C. M. (2003) Leadership style and power relations in quantity surveying in Hong Kong. *Construction Management and Economics*, **21**(8), 809–818.
42. Dainty, A. R. J., Bryman, A. and Price, A. D. F. (2002) Empowerment within the UK construction sector. *Leadership and Organizational Development Journal*, **23**(6), 333–342.
43. Fairholm (2004: 578).
44. Berry and Cartwright (2000: 347)
45. Cooper, C. D., Scandura, T. A. and Schriesheim, C. A. (2005) Looking forward but learning from our past: potential challenges developing authentic leadership theory and authentic leaders. *Leadership Quarterly*, **16**(3), 475–493; see also Toor, S. R. and Ofori, G. (2006) An integrated antecedental model of leadership development. *In*: Pietroforte, R. (Ed.) *Proceedings of the Joint international symposium of CIB W55/ W65/W86: Construction in the 21st century – local and global challenges*, 17–20 October 2006, Rome.
46. Chan, P. and Connolly, M. (2006) Combating skills shortages: examining careers advice in schools. *Construction Information Quarterly*, **8**(3), 131–137.
47. Handy (2001: 18).
48. Bloxham, T. (2005) *Keynote speech* delivered at the Emap radio sales conference, 25 May, Oulton Hall, Leeds.
49. Giddens (1984).

Chapter 3

1. Pearce, D. (2003) *The social and economic value of construction: the construction industry's contribution to sustainable development*. London: nCRISP.

2. See, for example, Mackenzie, S., Kilpatrick, A. R. and Akintoye, A. (2000) UK construction skills shortage response strategies and an analysis of industry perceptions. *Construction Management and Economics*, **18**, 853–862, and; Murray, M. and Langford, D. (2003) *Construction reports: 1944–98*. Edited text. Oxford: Blackwells.

3. Murray and Langford (2003: 213)

4. Pearce (2003); see also Meikle, J. and Dickson, M. (2006) Editorial: understanding the social and economic value of construction. *Building Research and Information*, **34**(3), 191–196.

5. Pearce (2003).

6. Ibid.

7. UK Government (1999: 8) *A better quality of life: a strategy for sustainable development for the UK*. London: HMSO.

8. Pearce, D. (2006: 201) Is the construction sector sustainable?: definitions and reflections. *Building Research and Information*, **34**(3), 201–207

9. Pearce (2003).

10. Turner, R. K. (2006) Sustainability auditing and assessment challenges. *Building Research and Information*, **34**(3), 197–200.

11. Pearce (2003).

12. Department for Business, Innovation and Skills (2009) *Construction statistics 2009*. Accessed through www.statistics.gov.uk on 21 January 2010.

13. Pearce (2003).

14. Ibid.

15. See Barker, K. (2004) *Delivering stability: securing our future housing needs*. UK: HM Treasury.

16. Olomolaiye, P. O., Jayawardane, A. K. W. and Harris, F. C. (1998) *Construction productivity management*. Englemere: CIOB.

17. See, for example, Ive, G., Gruneberg, S., Meikle, J. and Crosthwaite, D. (2004) *Measuring the competitiveness of the UK construction industry: volume 1*. November. London: HMSO, and; Blake, N., Croot, J. and Hastings, J. (2004) *Measuring the competitiveness of the UK construction industry: volume 2*. November. London: HMSO.

18. See, for example, Broadberry, S. and O'Mahony, M. (2004) Britain's productivity gap with the United States and Europe: a historical perspective, *National Institute Economic Review*, **189**, 72–85, and; Leitch, A. (2006) *Prosperity for all in the global economy: work class skills*. Final report. December. Norwich: HMSO.

19. Barrett, P. (2005) *Revaluing construction: a global CIB agenda*. Publication 305. Rotterdam: International council for research and innovation in building and construction (CIB).

20. Ive *et al.* (2004).

21. Blake *et al.* (2004).

22. See Broadberry and O'Mahony (2004).

23. See, for example, Griffith, R. and Harmgart, H. (2005) Retail productivity. *International Review of Retail, Distribution and Consumer Research*, **15**(3), 281–290, and; Griffith, R. and Harmgart, H. (2006) *How does UK retail productivity measure up?* AIM Executive briefing series. London: Advanced

Institute of Management Research (AIM); see also www.aimresearch.org for ongoing work on productivity.

24. Griffith and Harmgart (2006: 17).
25. Broadberry and O'Mahony (2004).
26. Blake *et al.* (2004).
27. Broadberry and O'Mahony (2004).
28. Gomolski, B. and Smith, M. (2007) *Gartner 2006–2007 IT spending and staffing report: North America*. Stamford, USA: Gartner Inc.
29. See, for example, Love, P. E. D., Irani, Z. and Edwards, D. J. (2004) Industry-centric benchmarking of information technology benefits, costs and risks for small-to-medium sized enterprises in construction. *Automation in construction*, **13**(4), 507–524.
30. See, for example, Blake *et al.* (2004), Ive *et al.* (2004), and Griffith and Harmgart (2005).
31. Crawford, P. and Vogl, B. (2006) Measuring productivity in the construction industry. *Building Research and Information*, **34**(3), 208–219.
32. Chan, P. and Kaka, A. (2004) Construction productivity measurement: a comparison of two case studies. *In*: Khosrowshahi, F. (Ed.) *Proceedings of the twentieth annual ARCOM conference*, 1–3 September 2004, Heriot-Watt University, Association of Researchers in Construction Management, v. 1, 3–12.
33. Radosavljevic, M. and Horner, R. M. W. (2002) The evidence of complex variability in construction labour productivity. *Construction Management and Economics*, **20**, 3–12.
34. Drewin, F. J. (1982) *Construction productivity*. New York: Elsevier Science.
35. Winch, G. and Carr, B. (2001) Benchmarking on-site productivity in France and the UK: a CALIBRE approach. *Construction Management and Economics*, **19**, 577–590.
36. See, for example, Thomas, H. R., Maloney, W. F., Horner, R. M. W., Smith, G. R. and Handa, V. K. (1990) Modelling construction labour productivity. *Journal of Construction Engineering and Management*, ASCE, **116**(4), 705–726.
37. Calvert, R. E., Bailey, G. and Coles, D. (1995) *Building management*, 6 Ed., Oxford: Butterworth-Heinemann.
38. Flanagan, R., Cattell, K. and Jewell, C. (2005) *Moving from construction productivity to construction competitiveness: measuring value not output*. Reading: University of Reading.
39. See, for example, Macarov, D (1982) *Worker productivity: myths and reality*. California: Sage Publications.
40. Chan, P. and Kaka, A. (2007: 244) The impacts of workforce integration on productivity. *In*: Dainty, A. R. J., Green, S. and Bagilhole, B. (Eds) *People and culture in construction: a reader*. London: Spons. Pp. 240–257.
41. Elias, J. and Scarbrough, H. (2004) Evaluating human capital: an exploratory study of management practice. *Human Resource Management Journal*, **14**(4), 21–40.
42. See www.accountingforpeople.gov.uk, and also Accounting for People (2003) *Accounting for people: report of the task force on human capital management*. London: AfP.

43. Becker, G. (1964; 1993) *Human capital: a theoretical and empirical analysis, with special reference to education.* 3rd Edn. Chicago: University of Chicago Press.
44. Grugulis, I. (2003a: 7) Putting skills to work: learning and employment at the start of the century. *Human Resource Management Journal*, **13**(2), 3–12.
45. Blenkinsopp, J. and Scurry, T. (2007) "Hey GRINGO!" the HR challenge of graduates in non-graduate occupations. *Personnel Review*, **36**(4), 623–637.
46. For a more detailed review of the critical issues specific to skills and the construction industry, see Chan, P. W. and Dainty, A. R. J. (2007) Resolving the UK construction skills crisis: a critical perspective on the research and policy agenda. *Construction Management and Economics*, **25**(4), 375–386.
47. Brown, A. W., Adams, J. D. and Amjad, A. A. (2007) The relationship between human capital and time performance in project management: a path analysis. *International Journal of Project Management*, **25**, 77–89.
48. See Mackenzie *et al.* (2000).
49. Winch, G. (1998) The growth of self-employment in British construction. *Construction Management and Economics*, **16**(5), 531–542.
50. Forde, C. and MacKenzie, R. (2007) Getting the mix right? The use of labour contract alternatives in UK construction. *Personnel Review*, **36**(4), 549–563.
51. Office of National Statistics (2008) *Construction Statistics 2008.* Newport: ONS.
52. Becker (1964).
53. Pearce (2006).
54. See Bloom, N., Conway, N., Mole, K., Möslein, K., Neely, A. and Frost, C. (2004) *Solving the Skills Gap: Summary Report from a CIHE/AIM Management Research Forum.* London: AIM, and; Groen, J. A. (2006) Occupation-specific human capital and local labour markets. *Oxford Economic Papers*, **58**(4), 722–741.
55. Groen (2006: 722).
56. Stevens, M. (1994) A theoretical model of on-the-job training with imperfect competition. *Oxford Economic Papers*, **46**, 537–562.
57. Groen (2006: 725).
58. Cockburn, C. (1983) *Brothers: male dominance and technological change.* London: Pluto Press.
59. Grugulis (2003a: 4).
60. Grugulis, I., Warhurst, C. and Keep, E. (2004) What's happening to 'skill'? *In:* Warhurst, C., Keep, E. and Grugulis, I. (Eds.) *The skills that matter.* Hampshire: Palgrave Macmillan. pp. 1–19.
61. Clarke, L. (1992) *The building labour process: problems of skills, training and employment in the British construction industry in the 1980s.* Occasional Paper No. 50, CIOB, Englemere.
62. Chan P. W. and Moehler, R. C. (2008) Construction skills development in the UK: transitioning between the formal and informal. *In:* Carter, K., Ogunlana, S. and Kaka, A. P. (Eds.): *Proceedings of the CIB W55/W65 joint conference: transforming through construction.* 15–17 November, Dubai, United Arab Emirates.
63. Becker (1964).
64. Groen (2006).

65. Dainty, A. R. J., Ison, S. G. and Root, D. S. (2005) Averting the construction skills crisis: a regional approach. *Local Economy,* **20**(1), 79–89.
66. Braverman, H. (1974) *Labor and monopoly capital: the degradation of work in the twentieth century.* New York: Monthly Review Press.
67. Beaney, W.D. (2006) *An exploration of sectoral mobility in the UK labour force: a principle components analysis.* Unpublished MSc thesis, Newcastle Business School, Northumbria University, UK.
68. Grugulis *et al.* (2004).
69. Becker (1964).
70. Clarke, L. and Herrmann, G. (2004) Cost vs. production: disparities in social housing construction in Britain and Germany. *Construction Management and Economics,* **22**, 521–532.
71. Grugulis (2003a: 5); see also Grugulis, I., Vincent, S. and Hebson, G. (2003) The rise of the 'network organisation' and the decline of discretion. *Human Resource Management Journal,* **13**(2), 45–59.
72. See, for example, Nyhan, B., Cressey, P., Tomassini, M., Kelleher, M. and Poell, R. (2004) European perspectives on the learning organisation. *Journal of European Industrial Training,* **28**(1), 67–92.
73. See, for example, Druker, J. and White, G. (1996) *Managing people in construction.* London: Institute of Personnel and Development, and; Druker, J., White, G., Hegewisch, A. and Mayne, L. (1996) Between hard and soft HRM: human resource management in the construction industry. Construction management and economics, **14**, 405–416.
74. Grugulis *et al.* (2004: 12).
75. Grugulis *et al.* (2004: 14).
76. Beckingsdale, T. and Dulaimi, M. F. (1997) *The Investors in People standard in UK construction organisations.* CIOB construction papers, 9–13.
77. Bell, E., Taylor, S. and Thorpe, R. (2002) A step in the right direction? Investors in People and the learning organisation. *British Journal of Management,* **13**, 161–171; see also Grugulis, I. and Bevitt, S. (2002) The impact of Investors in People: a case study of a hospital trust. *Human Resource Management Journal,* **12**(3), 44–60.
78. Elias and Scarbrough (2004), op. cit., p. 35.
79. See, for example, Callender, C. (1997) *Will NVQs work? Evidence from the construction industry.* Report by the Institute of Manpower Studies (IMS) for the Employment Department Group; Agapiou, A. (1998) A review of recent developments in construction operative training in the UK. *Construction Management and Economics,* **16**, 511–520, and; Grugulis, I. (2003b) The contribution of National Vocational Qualifications to the growth of skills in the UK. *British Journal of Industrial Relations,* **41**(3), 457–75.
80. Clarke, L. (2006) Valuing labour. *Building Research and Information,* **34**(3), 246–256.
81. Raidén, A. B. and Dainty, A. R. J. (2006) Human resource development in construction organisations: an example of a "chaordic" learning organisation? *The Learning Organisation,* **13**(1), 63–79; see also Chan, P. W. (2007) Managing human resources without Human Resources Management Departments: some exploratory

findings on a construction project. *In*: Hughes, W. (Ed.) *Proceedings of the CME25 Conference: Construction management and economics: past, present and future.* 16–18 July, University of Reading, UK.

82. Clarke, L. and Winch, C. (2004) Apprenticeship and applied theoretical knowledge. *Educational Philosophy and Theory*, **36**(5), 509–521.
83. Leitch (2006).
84. Broadberry and O'Mahony (2004).
85. Clarke and Herrmann (2004).
86. Forde and Mackenzie (2007).
87. Grugulis (2003a) and Grugulis *et al.* (2003).
88. Syben, G. (1998) A qualifications trap in the German construction industry: changing the production model and the consequences for the training system in the German construction industry. *Construction Management and Economics*, **16**, 593–601; for similar observations in the Australian context, see also Toner, P. (2006) Restructuring the Australian construction industry and workforce: implications for a sustainable labour supply. *The economic and labour relations review*, pp. 171–202.
89. *An inconvenient truth: a global warning* (2006) Paramount Classics (USA).
90. Pearce, D. W., Markandya, A. and Barbier, E. (1989) *Blueprint for a green economy.* London: Earthscan; see also Turner (2006), and; Turner, K. (2005) *The Blueprint Legacy: a review of Professor David Pearce's contribution to environmental economics and policy.* Norwich: Centre for Social and Economic Research on the Global Environment, University of East Anglia. Accessed through www.uea.ac.uk/env/ cserge/on 20 August 2007.
91. Turner (2006: 198)
92. Stern, N. (2006) *The economics of climate change.* Cambridge: Cambridge University Press.
93. Shipworth, D. (2007: 479) The Stern Review: implications for construction. *Building Research and Information*, **35**(4), 478–484.
94. Shipworth (2007: 482).
95. Pearce, D. W. and Atkinson, G. D. (1993: 103) Capital theory and the measurement of sustainable development: an indicator of "weak" sustainability. *Ecological Economics*, **8**, 103–108.
96. Devkota, S. R. (2005) Is strong sustainability operational? *Sustainable Development*, **13**, 297–310.
97. Pearce, D. W. and Barbier, E. (2000) *Blueprint for a sustainable economy.* London: Earthscan.
98. See Devkota (2005) and Turner (2006).
99. Shipworth (2007: 482).
100. Turner (2005).
101. Neumayer, E. (1999) Global warming: discounting is not the issue, but substitutability is. *Energy Policy*, **27**(1), 33–43.
102. Ekins, P. (2003) Identifying critical natural capital: conclusions about critical natural capital. *Ecological Economics*, **44**(2–3), 277–292.
103. Ibid.
104. Ekins (2003: 278).

105. Ibid.
106. See, for example, Turner, D. and Hartzell, L. (2004) The lack of clarity in the precautionary principle. *Environmental Values*, **13**, 449–460; Ekeli, K. (2004) Environmental risk, uncertainty and intergenerational ethics. *Environmental Values*, **13**, 421–448, and; Turner (2005).
107. Neumayer (1999) and Pearce (2006).
108. Pearce (2006).
109. Handy, C. (1994) *The empty raincoat.* UK: Random House.
110. England, R. W. (1998: 262) Should we pursue measurement of the natural capital stock? *Ecological Economics*, **27**, 257–266.
111. Wacknernagel, M., Monfreda, C., Schulz, N. B., Erb, K., Haberl, H. And Krausmann, F. (2004: 276) Calculating national and global ecological footprint time series: resolving conceptual challenges. *Land Use Policy*, **21**(3), 271–278.
112. Chiesura, A. and de Groot, R. (2003) Critical natural capital: a socio-cultural perspective. *Ecological Economics*, **44**(2–3), 219–231.
113. Chiesura and de Groot (2003: 224).
114. Spash, C. L. (2007) The economics of climate change impacts à la Stern: novel and nuanced or rhetorically restricted. *Ecological Economics*, **63**(4), 706–713.
115. Pearce (2006).
116. See, for example, Samuelson, P. A. (1947) *Foundations of economic analysis.* Cambridge, USA: Harvard University Press, and; van Praag, B. M. S. (2007) Perspectives from the happiness literature and the role of new instruments for policy analysis. *Economic Studies*, **53**(1), 42–68.
117. Neumayer (1999: 41).
118. Stern (2006).
119. Wacknernagel *et al.* (2004: 272).
120. Herring, H. (2006: 16) Energy efficiency: a critical view. *Energy*, **31**(1), 10–20.
121. Herring (2006: 19).
122. England (1998: 264).
123. Stern (2006).
124. See also Shipworth (2007); Neumayer (1999), and; Neumayer, E. (2003) *Weak versus strong sustainability: exploring the limits of two opposing paradigms.* 2nd Edn. Cheltenham: Edward Elgar.
125. Stern (2006).
126. Spash (2007: 712).
127. Turner (2005).
128. Hughes, W., Ancell, D. Gruneberg, S. and Hirst, L. (2004) Exposing the myth of the 1:5:200 ratio relating initial cost, maintenance and staffing costs of office buildings. *In:* Khosrowshahi, F. (Ed.) *Proceedings of the twentieth ARCOM conference,* Association of researchers in construction management, 1–3 September, Heriot-Watt University, Edinburgh, 373–381.
129. Ive, G. (2006) Re-examining the costs and value ratios of owning and occupying buildings. *Building Research and Information*, **34**(3), 230–245.
130. Ive (2006: 241).
131. See, for example, Lowe, R. (2007) Assessing the challenges of climate change for the built environment. *Building Research and Information*, **35**(4), 343–350.

132. Banfill, P. F. G. and Peacock, A. D. (2007) Energy-efficient new housing: the UK reaches for sustainability. *Building Research and Information*, **35**(4), 426–436.
133. Boardman, B., Darby, S., Killip, G., Hinnells, M., Jardine, C. N., Palmer, J. and Sinden, G. (2005: 11) *40% house*. February. Oxford: Environmental change institute, University of Oxford.
134. Banfill and Peacock (2007).
135. See also Wacknernagel *et al.* (2004); Herring (2006), and; Spash (2007).
136. Schiller, G. (2007) Urban infrastructure: challenges for resource efficiency in the building stock. *Building Research and Information*, **35**(4), 399–411.
137. See Meikle and Dickson (2006).
138. Boardman *et al.* (2005).
139. Lowe (2007).
140. Banfill and Peacock (2007).
141. Lowe (2007), op. cit., p. 347.
142. Ibid.; see also Hamza, N. A. and Greenwood, D. J. (2009) Energy conservation regulations: impacts on design and procurement of low energy buildings. *Building and Environment*, **44**(5), 929–936.
143. See, for example, Gann, D. M., Wang, Y. and Hawkins, R. (1998) Do regulations encourage innovation? the case of energy efficiency in housing. *Building Research and Information*, **26**(5), 280–296, and; Meacham, B., Bowen, R., Traw, J. and Moore, A. (2005) Performance-based building regulation: current situation and future needs. *Building Research and Information*, **33**(2), 91–106.
144. Chappells H. and Shove E. (2004) *Report on the 'Future Comforts' workshop*, www.lancs.ac.uk/fass/sociology//research/projects/futcom/documents/webreport.htm accessed on 23 August 2007.
145. Wood G. and Newborough M. (2003) Dynamic energy consumption indicators for domestic appliances: environment, behaviour and design. *Energy and Buildings*, **35** (8), 821–841.
146. Pett J. and Guertler P. (2004) *User behaviour in energy efficient homes*. Report by the Association for the Conservation of Energy for the Energy Saving Trust and the Housing Corporation. www.ukace.org/research/behaviour/User%20Behaviour% 20-%20Phase%202%20report%20v1.0.pdf accessed on 23 August 2007.
147. Ibid.
148. Wood and Newborough (2003).
149. Banfill and Peacock (2007).
150. Boardman *et al.* (2005).
151. *How We Built Britain – The North: full steam ahead* (2007) BBC (UK).
152. Bridger, J. C. and Luloff, A. E. (2001: 459–460) Building the sustainable community: is social capital the answer? *Sociological Inquiry*, **71**(4), 458–472; this trend of shifting power away from the political elite to community actors at the grassroots level is also encapsulated in the contemporary aspirations of the coalition government in the UK of building a "Big society".
153. Ibid.
154. Girardet, H. (1999) *Creating sustainable cities*. Totnes: Green Books/The Schumacher Society.

155. Rogers, R. (1997) *Cities for a small planet*. London: Faber and Faber; see also Urban Task Force (1999) *Towards an urban renaissance*. London: E&FN Spon.
156. Doughty, M. R. C. and Hammond, G. P. (2004) Sustainability and the built environment at and beyond the city scale. *Building and Environment*, **39**, 1223–1233.
157. Ibid.
158. Office of the Deputy Prime Minister (2005) *Sustainable communities: homes for all*. London: ODPM.
159. Adapted from: Bridger and Luloff (2001); Doughty and Hammond (2004), and; ODPM (2005)
160. Bridger and Luloff (2001: 460).
161. Bridger and Luloff (2001: 458).
162. Bridger and Luloff (2001: 463).
163. See, for example, Zeisel, J. (2003) Marketing therapeutic environments for Alzheimer's care. *Journal of Architectural and Planning Research*, **20**(1), 75–86, and; Ulrich, R. Quan, X. Zimring, C., Joseph, A. and Choudhary, R. (2004) *The role of the physical environment in the hospital of the 21ˢᵗ century: a once-in-a-lifetime opportunity*. Report, Center for health systems and design, College of Architecture, Texas A & M University.
164. Whyte, J. K. and Gann, D. M. (2003) Design quality indicators: work in progress. *Building Research and Information*, **31**(5), 387–398.
165. Ibid.
166. See, for example, Rouse, J. (2004) Measuring value or only cost: the need for new valuation methods. *In*: Macmillan, S. (Ed.) *Designing better buildings*. London: E & FN Spon. 55–71, and; Macmillan, S. (2006) Added value of good design. *Building Research and Information*, **34**(3), 257–271.
167. Abdul Samad, Z. and Macmillan, S. (2005) The valuation of intangibles: explored through primary school design. *In*: Emmitt, S. (Ed.) *Proceedings of the CIB W096 conference on "Designing value: new directions in architectural management*. 2–4 November, Lyngby, Denmark, Technical University of Denmark.
168. Bridger and Luloff (2001).
169. Macmillan (2006).
170. Giddings, B. and Hopwood, B. (2006: 346–347) From evangelist bureaucrat to visionary developer: the changing character of the Master Plan in Britain. *Planning, Practice and Research*, **21**(3), 337–348.
171. Veenstra, G. (2002) Explicating social capital: trust and participation in the civil space. *Canadian Journal of Sociology*, **27**(4), 547–572.
172. Coleman, J. S. (1988: S100–S101) Social capital in the creation of human capital. *The American Journal of Sociology*, **94**, Supplement: Organisations and institutions: sociological and economic approaches to the analysis of social structure, S95–S120.
173. See, for example, Coase, R. (1937): The nature of the firm. *Economica*, **4**, 386–405, and; Williamson, O. E. (1981) The economics of organization: the transaction cost approach. *American Journal of Sociology*, **87**(3), 548–575.
174. Swan, W., McDermott, P., Wood, G., Thomas, A., and Abbott, C. (2002) *Trust in construction: achieving cultural change*. Manchester: Centre for Construction Innovation in the Northwest.

175. Rousseau, D. M., Sitkin, S. B., Burt, R. S. and Camerer, C. (1998) Not so different after all: a cross-discipline view of trust. *Academy of Management Review*, **23**(3), 393–404.

176. Smyth, H. (2003) *Developing client-contractor trust: a conceptual framework for management in project working environments.* London: University College London.

177. See Latham, M. (1993) *Trust and money.* London: HMSO, and; Latham, M. (1994) *Constructing the team.* London: HMSO.

178. Poppo, L. and Zenger, T. (2002: 721) Do formal contracts and relational governance function as substitutes or complements? *Strategic Management Journal*, **23**, 707–725.

179. Coleman (1988: S101).

180. Bridger and Luloff (2001: 467).

181. See, for example, Chan, P. W. and Räisänen, C. (2009) Informality and emergence in construction.

182. *Construction Management and Economics*, **27**(10), 907–912.

183. Giddens, A. (1984) *The constitution of society.* Cambridge: Polity Press.

184. Spangenberg, J. H. (2001: 184) Investing in sustainable development: the reproduction of manmade, human, natural and social capital. *International Journal of Sustainable Development*, **4**(2), 184–201.

Chapter 4

1. We draw particular inspiration here from the field of organisational institutionalism. Therefore, we seek to broaden the view of institutions beyond conventional governance and public bodies, to include other societal actors such as corporate firms and community actors. This forms the thrust of this chapter, where we examine both theoretical perspectives and practical insights from our interviews to discuss the interactions across these three broad categories of actors to explore areas for joined-up thinking and working. Our thinking is informed by inter alia: Greenwood, R., Oliver, C., Suddaby, R. and Sahlin-Andersson, K. (Eds.) (2008) *The SAGE handbook of organizational institutionalism.* London: Sage.

2. For further reading on institutional perspectives of alignment, the reader is referred to two seminal papers, which later formed the basis for the field of organisational institutionalism. These include Meyer, J.W. and Rowan, B. (1977). Institutionalized organizations: formal structure as myth and ceremony. *American Journal of Sociology*, **83**(2), 340–363, and; DiMaggio, P.J. and Powell, W.W. (1983) The iron cage revisited: institutional isomorphism and collective rationality in organizational fields. *American Sociological Review*, **48**(2), 147–160.

3. Dodgson, M., Gann, D. M. and Salter, A. (2008) *The management of technological innovation: strategy and practice.* 2nd Edn. Oxford: Oxford University Press.

4. Putnam, R. D. (2000) *Bowling alone: the collapse and revival of American community.* New York: Simon and Schuster.

5. Sobel, J. (2002) Can we trust social capital? *Journal of economic literature*, **40**(1), 139 – 154.

6. Peters, B. G. and Pierre, J. (1998) Governance without government? Rethinking public administration. *Journal of Public Administration Research and Theory*, **8**(2), 223–243.

7. Talbot, C. (2004: 111) Executive agencies: have they improved management in government? *Public Money and Management*, **24**(2), 104–112.

8. See, for example, Martin, A. and Scott, I. (2003) The effectiveness of the UK landfill tax. *Journal of Environmental Planning and Management*, **46**(5), 673–689, and; Revell, A. and Blackburn, R. (2005) The business case for sustainability? An examination of small firms in the UK's construction and restaurant sectors. *Business Strategy and the Environment*, **16**(6), 404–420.

9. Hansford, A., Hasseldine, J. and Woodward, T. (2004: 208) The UK climate change levy: good intentions but potentially damaging for business. *Corporate Social Responsibility and Environmental Management*, **11**(4), 196–210.

10. Peters and Pierre (1998).

11. Peters and Pierre (1998: 223).

12. Peters and Pierre (1998: 225).

13. See also Gospel, H. and Foreman, J. (2006) Inter-firm training co-ordination in Britain. *British Journal of Industrial Relations*, **44**(2), 191–214.

14. Mulgan, G. (2005a) *Joined-up government: past, present and future*. Bristol: The Young. Accessed through www.youngfoundation.org.uk on 28 August 2007.

15. Mulgan, G. (2005b) Lessons from government. *Politics and policy lecture*. March. Bristol: The Young Foundation. Accessed through www.youngfoundation.org.uk on 28 August 2007.

16. Peters and Pierre (1998)

17. Caporaso, J. A. and Wittenbrinck, J. (2006: 472) The new modes of governance and political authority in Europe. *Journal of European Public Policy*, **13**(4), 471–480.

18. Caporaso and Wittenbrinck (2006: 474).

19. See, for example, Dolowitz, D. P. and Marsh, D. (2000) Learning from abroad: the role of policy transfer in contemporary policy-making. *Governance: an International Journal Of Policy and Administration*, **13**(1), 5–24, and; Mossberger, K. and Wolman, H. (2003) Policy transfer as a form of prospective policy evaluation: challenges and recommendations. *Public Administration Review*, **63**(4), 428–440.

20. Peters and Pierre (1998: 225).

21. Handy, C. (2001) *The elephant and the flea: looking backwards to the future*. London: Hutchinson.

22. Child, J. and Rodrigues, S. B. (2003) Corporate governance and new organizational forms: issues of double and multiple agency. *Journal of Management and Governance*, **7**(4), 337–360.

23. Shleifer, A. and Vishny, R. W. (1997: 737) A survey of corporate governance. *The Journal of Finance*, **52**(2), 737–783.

24. Denis, D. K. (2001) Twenty-five years of corporate governance research... and counting. *Review of Financial Economics*, **10**(3), 191–212.

25. See, for example, Mintzberg, H. (1973) *The nature of managerial work*. New York: Harper and Row.

26. See, for example, Stinchcombe, A.L. (1959) Bureaucratic and craft administration of production: a comparative study. *Administrative Science Quarterly*, **4**, 168–187; Stinchcombe, A.L. (1985) Contracts as hierarchical documents. *In*: Stinchcombe, A.L. and Heimer, C.A. (Eds) *Organization theory and project management: administering uncertainty in Norwegian offshore oil*, Oslo: Norwegian University Press, 121–171, and; Stinchcombe, A. (1990) *Information and organizations*. Berkeley, CA: University of California Press.

27. Kahneman, D. and Tversky, A. (1982) On the study of statistical intuitions. *Cognition*, March, **11**(2), 123–141.

28. Hart, O. (1995) Corporate governance: some theory and implications. *The Economic Journal*, **105**(430), 678–689.

29. See, for example, Coase, R. (1937). The nature of the firm. *Economica*, **4**(4), 386–405; Williamson, O. (1985). *The economic institutions of capitalism*. New York: Free Press, and; Williamson, O. E. (1998) The institutions of governance. *The American Economic Review*, **88**(2), 75–79.

30. Hart (1995: 678).

31. See Hart (1995); Pietroforte, R. (1997) Communication and governance in the building process. *Construction Management and Economics*, **15**(1), 71–82, and; Winch, G. M. (2001) Governing the project process: a conceptual framework. *Construction Management and Economics*, **19**(8), 799–808.

32. Chan, P. W. and Räisänen, C. (2009) Informality and emergence in construction. *Construction Management and Economics*, **27**(10), 907–912.

33. Nelson, R. R. and Winter, S. G. (1982) *An evolutionary theory of economic change*. Cambridge, MA: Belknap Press.

34. Styhre, A. (2004) Rethinking knowledge: A bergsonian critique of the notion of tacit knowledge. *British Journal of Management*, **15**, 177–188.

35. Shankman, N. A. (1999) Reframing the debate between agency and stakeholder theories of the firm. *Journal of Business Ethics*, **19**, 319–334.

36. See, for example, Piore, M. J. and Sabel, C. (1984) *The second industrial divide: possibilities for prosperity*. New York: Basic Books; Sako, M. (1992) *Price, quality and trust: inter-firm relations in Britain and Japan*. Cambridge: Cambridge University Press; Putnam, R. D. (1993) *Making democracy work: civic traditions in modern Italy*. Princeton: Princeton University Press; Fukuyama, F. (1995) *The social virtues and the creation of prosperity*. New York: Free Press; Rousseau, D. M., Sitkin, S. B., Burt, R. S. and Camerer, C. (1998) Not so different after all: a cross-discipline view of trust. *Academy of Management Review*, **23**(3), 393–404, and; Adler, P. (2001) Market, hierarchy, and trust: the knowledge economy and the future of capitalism. *Organisation Science*, **12**(2), 215–234.

37. Child and Rodrigues (2003).

38. Fox, A. (1966) *Industrial sociology and industrial relations*. Royal Commission research paper no. 3. London: HMSO.

39. Child and Rodrigues (2003: 340).

40. See, for example, Whittington, R. (1996) Strategy as practice. *Long Range Planning*, **29**, 731–735, and; Jarzabkowski, P. (2004) Strategy as practice: recursiveness, adaptation, and practices-in-use. *Organization Studies*, **25**(4), 529–560.

41. Denis (2001).
42. Child and Rodrigues (2003).
43. Claessens, S. (2006) Corporate governance and development. *The World Bank Research Observer*, **21**(1), 91–122.
44. Branston, J. R., Cowling, K. and Sugden, R. (2006: 190) Corporate governance and the public interest. *International Review of Applied Economics*, **20**(2), 189–212.
45. Aguilera, R. V., Williams, C. A., Conley, J. M. and Rupp, D. E. (2006) Corporate governance and social responsibility: a comparative analysis of the UK and the US. *Corporate Governance*, **14**(3), 147–158.
46. World Business Council for Sustainable Development (1999) *Corporate social responsibility*. WBCSD; cf. Moir, L. (2001) What do we mean by corporate social responsibility? *Corporate Governance*, **1**(2), 16–22.
47. See, for example, Friedman, M. (1962) *Capitalism and freedom*. Chicago: University of Chicago Press.
48. McWilliams, A., Siegel, D. S. and Wright, P. M. (2006) Corporate social responsibility: strategic implications. *Journal of Management Studies*, **43**(1), 1–18.
49. Bowles, S. and Gintis, H. (2002: F421) Social capital and community governance. *The Economic Journal*, **112**, F419–F436.
50. Gladwell, M. (2000) *The tipping point: how little things can make a big difference*. London: Little, Brown and Company.
51. Bowles and Gintis (2002).
52. Putnam, R. D. (2007: 138) *E Pluribus Unum*: Diversity and community in the twenty-first century the 2006 Johan Skytte Prize lecture. *Scandinavian Political Studies*, **30**(2), 137–174.
53. Bowles and Gintis (2002: 433).
54. Sullivan, H. (2001) Modernisation, democratisation and community governance. *Local Government Studies*, **27**(3), 1–24.
55. Sullivan (2001: 8).
56. Sommerville, P. (2005: 123) Community governance and democracy. *Policy and Politics*, **33**(1), 117–144.
57. Bridger and Luloff (2001).
58. Mulgan (2005b).
59. Putnam (2007).
60. Sobel (2002).
61. Bowles and Gintis (2002: F430–F431).
62. Dunleavy, P., Margetts, H., Bastow, S. and Tinkler, J. (2005) New public management is dead: long live digital-era governance. *Journal of Public Administration Research and Theory*, **16**, 467–494.
63. Mulgan (2005a).
64. See, for example, Chesborough (2003) *Open innovation: the new imperative for creating and profiting from technology*. Boston: Harvard Business School; Dodgson *et al.*, 2005, and; von Hippel, E. (2005) *Democratizing innovation*. Cambridge, USA: MIT Press. These references make the case for harnessing technological advancements to open up the innovation process.

65. See, for example, Department for Education and Employment (2000) *An assessment of skill needs in construction and related Industries*, Skill Dialogues: Listening to Employers Research Papers, London: DfEE and Business Strategies Ltd; Dainty, A. R. J. and Edwards, D. J. (2003) The UK building education recruitment crisis: a call for action. *Construction Management and Economics*, **21**, 767–775; Agapiou, A. (2002) Perceptions of gender roles and attitudes toward work among male and female operatives in the Scottish construction industry. *Construction Management and Economics*, **20**, 697–705, and; Chan, P. W. and Dainty, A. R. J. (2007) Resolving the UK construction skills crisis: a critical perspective on the research and policy agenda. *Construction Management and Economics*, **25**(4), 375–386.

66. Leitch, A. (2006) *Prosperity for all in the global economy: work class skills.* Final report. December. Norwich: HMSO.

67. See, for example, Grugulis, I. (2007) *Skills, training and human resource development: a critical text.* Basingstoke: Palgrave Macmillan.

68. See, for example, Gospel and Foreman (2006), and; Chan, P. and Moehler, R. (2007) Developing a 'road-map' to facilitate employers' role in engaging with the skills development agenda. *In*: Boyd, D. (Ed.) *Proceedings of the twenty-third ARCOM conference*, 3–5 September 2007, Belfast, UK, Association of Researchers in Construction Management, **1**, 409–418.

69. Tzortzopoulos, P., Cooper, R., Chan, P. and Kagioglou, M. (2006) Clients' activities at the design front-end. *Design Studies*, **27**(6), 657–683.

70. See, for example, Devlin, A. S., and Arneill, A. B. (2003) Health care environments and patient outcomes: a review of the literature. *Environment and Behaviour*, **35**(5), 665–694, and; Gesler, W. B. M., Curtis, S., Hubbard, P. and Francis, S. (2004) Therapy by design: evaluating the UK hospital building programme. *Health and Place*, **10**, 117–128.

71. After United Nations Development Programme (1997) *Governance for sustainable human development.* New York: United Nations; Van Wyk, L. and Chege, L. (2004) Globalisation, corporate governance and the construction industry. *In*: Charoenngam, C. and Ogunlana, S. O. (Eds.) *Proceedings of International Symposium on globalisation and construction: meeting the challenges, reaping the benefits.* 17–19 November. Asian Institute of Technology, Thailand. Accessed through www.buildnet.co.za/akani/2005/mar/pdfs/akani_contractor.pdf on 27 August 2007, and; Kaufmann, D. (2005) Myths and realities of governance and corruption. *In*: World Economic Forum *Global competitiveness report 2005–2006.* New York: Oxford University Press. pp. 81–95.

72. Castells, M. (1996) *The rise of the network society.* Oxford: Blackwell, and; Castells, M. (2002) Internet and the network enterprise. Plenary address given to the *18th EGOS (European Group for Organizational Studies) Colloquium*, Barcelona, 4 July.

73. Murray and Langford (2003).

74. Kerr, D. (1998: 2283–2284) The Private Finance Initiative and the changing governance of the built environment. *Urban Studies*, **35**(12), 2277–2301.

75. Ibid.

76. Lansley, P. R. (1987: 148) Corporate strategy and survival in the UK construction industry. *Construction Management and Economics*, **5**(2), 141–155.
77. Betts, J. and Holden, R. (2003) Organisational learning in a public sector organisation: a case study in muddled thinking. *Journal of Workplace Learning*, **15**(6), 280–287.
78. See, for example, Tzortzopoulos, P., Cooper, R., Chan, P. and Kagioglou, M. (2006) Clients' activities at the design front-end. *Design Studies*, **27**(6), 657–683.
79. Rose, N. (1999) *Powers of freedom: reframing political thought.* Cambridge: Cambridge University Press.
80. Lansley (1987).
81. Burnham, P. (2001: 128–129) New Labour and the politics of depoliticisation. *British Journal of Politics and International Relations*, **3**(2), 127–149.
82. Minton, A. (2006) *The privatisation of public space: what kind of world are we building?* March. London: RICS.
83. Barry, A. (2002: 271) The anti-political economy. *Economy and Society*, **31**(2), 268–284.
84. González, M., Arruñada, B. and Fernández, A. (1998: 433) Regulation as a cause of firm fragmentation: the case of the Spanish construction industry. *International Review of Law and Economics*, **18**, 433–450.
85. Ministerio de Trabajo y Asuntos Sociales España (2006) *Ley 32/2006 reguladora de la subcontratación en el Sector de la Construcción.* Madrid: MTAS. Accessed through www.mtas.es/Insht/Legislation/L/L32subcont.htm on 24 March 2008.
86. See, for example, Lane, C. (1988) Industrial change in Europe: the pursuit of flexible specialisation in Britain and West Germany. *Work, Employment and Society*, **2**(2), 141–168; Clarke, L. and Wall, C. (1996) *Skills and the construction process: a comparative study of vocational training and quality in social housebuilding.* The Policy Press, Housing and Construction Industry Research Programme; Broadberry, S. and O'Mahony, M. (2004) Britain's productivity gap with the United States and Europe: a historical perspective, *National Institute Economic Review*, **189**, 72–85, and; Leitch (2006).
87. Bartram, D. (2004) Labor migration policy and the governance of the construction industry in Israel and Japan. *Politics and Society*, **32**(2), 131–170.
88. Chan, P., Clarke, L. and Dainty, A. (forthcoming) The dynamics of migrant employment in construction: can supply of skilled labour ever match demand? *In:* Ruhs, M. and Anderson, B. (Eds.) *Who needs migrant workers? Labour shortages, immigration, and public policy.* Oxford: Oxford University Press.
89. Harris, F., McCaffer, R. and Edum-Fotwe, F. (2006) *Modern construction management.* 6th Edn. Oxford: Wiley-Blackwell.
90. See, for example, Hillebrandt, P. M. (2000) *Economy theory and the construction industry.* 3 Ed. Hampshire: Palgrave Macmillan, and; Ive, G. J. and Gruneberg, S. L. (2000) *The economics of the modern construction sector.* Basingstoke: Macmillan.
91. Rebeiz, K. S. and Salameh, Z. (2006) Relationship between governance structure and financial performance in construction. *Journal of Management in Engineering*, **22**(1), 20–26.
92. The Chartered Institute of Building (2006) *Corruption in the UK construction industry: survey 2006.* Englemere: CIOB.

93. Pearce, D. (2003) *The social and economic value of construction: the construction industry's contribution to sustainable development.* London: nCRISP.
94. Groák, S. (1992) *The idea of building: thought and action in the design and production of buildings.* Oxon: Taylor and Francis.
95. Groák, S (1994: 291) Is construction an industry? Notes towards a greater analytic emphasis on external linkages. *Construction Management and Economics*, **12**, 287–293.
96. Groák (1994: 288).
97. Groák (1994: 291).
98. Phelps-Brown, E. H. (1968: 170–171) *Report of the committee of inquiry under Professor E H Phelps Brown into certain matters concerning labour in building and civil engineering.* London: HMSO.
99. See also Barrett, P. (2008) *Revaluing construction.* Oxford: Blackwell.
100. Eccles, R. G. (1981) The quasi-firm in the construction industry. *Journal of Economic Behaviour and Organisation*, **5**, 335–357.
101. Pryke, S. (2004: 790) Analysing construction project coalitions: exploring the application of social network analysis. *Construction Management and Economics*, **22**(8), 787–797.
102. See, for example, Rubery, J., Carroll, C., Cooke, FL., Grugulis, I. and Earnshaw, J. (2004) Human resource management and the permeable organization: the case of the multi-client call centre. *Journal of Management Studies*, **41**(7), 1199–1222, and; Marchington, M., Grimshaw, D., Rubery, J. and Willmott, H. (2005) (Eds.) *Fragmenting work: blurring organizational boundaries and disordering hierarchies.* Oxford: Oxford University Press.
103. See, for example, Grugulis, I., Vincent, S. and Hebson, G. (2003) The rise of the 'network organisation' and the decline of discretion. *Human Resource Management Journal*, **13**(2), 45–59; Kinnie, N. J., Swart, J. and Purcell, J. (2005) Influences on the choice of HR system: the network organisation perspective. *International Journal of Human Resource Management*, **16**(6), 1004–1028, and; Grimshaw, D. and Rubery, J. (2005) Inter-capital relations and the network organisation: redefining the work and employment nexus. *Cambridge Journal of Economics*, **29**(6), 1027–1051.
104. See, for example, Hobday, M. (2000) The project-based organisation: an ideal form for managing complex products and systems? *Research Policy*, **29**(7/8), 871–894; Whitley, R. (2006) Project-based firms: new organizational form or variations on a theme., *Industrial and Corporate Change*, **15**(1), 77–99, and; Bredin, K. and Söderlund, J. (2007) Reconceptualising line management in project-based organisations: the case of competence coaches at Tetra Pak. *Personnel Review*, **36**(5), 815–833.
105. Marchington *et al.* (2005).
106. Olympic Development Authority (2007) *Procurement policy: executive summary.* March. London: London 2012.
107. Rubery *et al.* (2004).
108. Marchington, M. and Vincent, S. (2004) Analysing the influence of institutional, organizational and interpersonal forces in shaping inter-organizational relations. *Journal of Management Studies*, **41**(6), 1029–1056.

109. Marchington, M., Carroll, M., Pass, S., Grimshaw, D. and Rubery, J. (2009) *Managing people in networked organizations: research into practice.* London: Chartered Institute of Personnel and Development.
110. Winch (2001: 805–806).
111. See, for example, Proverbs, D. G. and Holt, G. D. (2000) Reducing construction costs: European best practice supply chain implications. *European Journal of Purchasing and Supply Management,* **6** 149–158, and; Vrijhoef, R. and Koskela, L. (2000) The four roles of supply chain management in construction. *European Journal of Purchasing and Supply Management,* **6**, 169–178.
112. See, for example, Bresnen, M. and Marshall, N. (2001) Understanding the diffusion and application of new management ideas in construction. *Engineering, Construction and Architectural Management,* **8**(5/6), 335–345, and; Briscoe, G. and Dainty, A. (2005) Construction supply chain integration: an elusive goal? *Supply Chain Management: an International Journal,* **10**(4), 319–326.
113. London, K. A. and Kenley, R. (2001: 786) An industrial organisation economic supply chain approach for the construction industry: a review. *Construction Management and Economics,* **19**(8), 777–788.
114. Pryke (2004) and Pryke, S. (2005) Towards a social network theory of project governance. *Construction Management and Economics,* **23**(9), 927–939.
115. Nohria, N. and Eccles, R.G. (Eds) (1992) *Networks and organisations.* Boston, MA: Harvard Business School Press.
116. Pryke (2004: 795).
117. Swan, W., McDermott, P., Wood, G., Thomas, A., and Abbott, C. (2002) *Trust in construction: achieving cultural change.* Manchester: Centre for Construction Innovation in the Northwest.
118. See, for example, Dodgson *et al.* (2005).
119. Harland, C. M. (1996) Supply chain management: relationships, chains and networks. *British Journal of Management,* **7**, Special issue on supply chain management, S63–S80.
120. Van de Ven, A. (2007) *Engaged scholarship: a guide for organizational and social research.* Oxford: Oxford University Press.
121. Pryke, S. and Pearson, S. (2007) An analytical Anglo-French comparative study of construction procurement and management strategies. *RICS Research Paper Series,* **7**(5), July. London: RICS.
122. See, for example, Murray, M. D., Tookey, J. E. and Chan, P. (2001) Respect for people: looking at KPIs through 'younger eyes'! *In:* Akintoye, A. (Ed.) *Proceedings of the seventeenth annual ARCOM conference,* 5–7 September 2001, University of Salford, Association of Researchers in Construction Management, v. 2, 671–681, and; Moore, D. R. (2001) William the Sen to Bob the Builder: non-cognate cultural perceptions of constructors. *Engineering, Construction and Architectural Management,* **8**(3), 177–184.
123. Murray *et al.* (2001), and; Langford, D. A. and Robson, P. (2003) The representation of the professions in the cinema: the case of construction engineers and lawyers. *Construction Management and Economics,* **21**, 799–807.

124. MORI (1998) *Children's attitudes towards the construction industry: a research study among 11–16 year-olds.* Market and Opinion Research International. Bircham Newton: CITB.
125. Moore (2001).
126. See also Dainty, A. R. J., Bagilhole, B. M. and Neale, R. H. (2000) A grounded theory of women's career under-achievement in large UK construction companies. *Construction Management and Economics,* **18**, 239–250.
127. Langford and Robson (2003: 804).
128. Moore (2001).
129. Department for Education and Skills (2003) *The future of higher education.* January. Norwich: HMSO.
130. Tomlinson, M. (2004) *14–19 Curriculum and qualifications reform: interim report of the working group on 14–19 reform.* Nottinghamshire: Department for Education and Skills (DfES).
131. See Leitch (2006), and; Department for Education and Skills (2005) *14–19 Education and skills.* February. Norwich: HMSO.
132. See, for example, Clarke, L. (1999) The changing structure and significance of apprenticeship with special reference to construction. *In*: Ainley, P and Rainbird, H (Eds) *The nature of apprenticeship.* London: Kogan Page. pp. 25–40, and; Clarke, L. and Winch, C. (2004) Apprenticeship and applied theoretical knowledge. *Educational Philosophy and Theory,* **36**(5), 509–521.
133. Clarke, L. and Herrmann, H. (2007) Divergent divisions of labour. *In*: Dainty, A. R. J., Green, S. and Bagilhole, B. (Eds) *People and culture in construction: a reader.* London: Spons. pp. 85–105.
134. Woudhuysen, J. and Abley, I. (2004) *Why is construction so backward?* Chichester: John Wiley and Sons.
135. See, for example, Delargy, M. (2001) At boiling point. *Building,* 26 January, 24–27.
136. Moore (2001).
137. See, for example, Dainty *et al.* (2000), and; Commission for Architecture and the Built Environment (2005) *Black and minority ethnic representation in the built environment professions.* London: CABE/Royal Holloway University of London.
138. Gale, A. W. and Davidson, M. (2006) *Managing diversity and equality in construction: initiatives and practice.* London: Taylor and Francis.
139. Winch, G. M. (2000: 143) Institutional reform in British construction: partnering and private finance. *Building Research and Information,* **28**(2), 141–155.
140. Clarke and Herrmann (2007: 92–93).
141. Murray and Langford (2003).
142. Pryke and Pearson (2007: 41, 43).
143. Ibid.
144. For a study that shares many of Pryke and Pearson's (2007) conclusions, the reader is referred to Anglo–German comparisons explained in Clarke, L. and Herrmann, G. (2004) Cost vs. production: disparities in social housing construction in Britain and Germany. *Construction Management and Economics,* **22**, 521–532. Further compelling evidence from an Anglo–French comparative perspective can be found in

Winch, G. and Carr, B. (2001) Benchmarking on-site productivity in France and Great Britain: a CALIBRE approach. *Construction Management and Economics,* **19**(6), 577–590.

145. Clarke and Herrmann (2007: 92).

146. Ibid.

147. Clarke and Herrmann (2007: 93).

148. Chan, P. and Kaka, A. (2007) The impacts of workforce integration on productivity. In: Dainty, A. R. J., Green, S. and Bagilhole, B. (Eds) *People and culture in construction: a reader.* London: Spons. pp. 240–257.

149. Green, S. (1999) The missing arguments of lean construction. *Construction Management and Economics,* **17**, 133–137.

150. Druker, J., White, G., Hegewisch, A. and Mayne, L. (1996) Between hard and soft HRM: HRM in the construction industry. *Construction Management and Economics,* **14**, 405–416.

151. See, for example, Beckingsdale, T. and Dulaimi, M. F. (1997) The Investors in People standard in UK construction organisations. *CIOB construction papers,* 9–13; Forde, C. and MacKenzie, R. (2004) Cementing skills: training and labour use in UK construction. *Human Resource Management Journal,* **14**(3), 74–88, and; Clarke, L. (2006) Valuing labour. *Building Research and Information,* **34**(3), 246–256.

152. Seymour, D. and Rooke, J. (1995: 513) The culture of the industry and the culture of research. *Construction Management and Economics,* **13**, 511–523.

153. Seymour and Rooke (1995: 521).

154. Seymour and Rooke (1995: 522).

155. Foxell, S. (Ed.) (2003) *The professionals' choice: the future of the built environment professions.* London: CABE and RIBA.

156. Davies, W. and Knell, J. (2003: 135) Conclusion. In: Foxell, S. (Ed.) *The professionals' choice: the future of the built environment professions.* London: CABE and RIBA. 131–141.

157. Jobling, A. (2003) Regulatory scenario. In: Foxell, S. (Ed.) *The professionals' choice: the future of the built environment professions.* London: CABE and RIBA. 37–54.

158. Curry, A. and Howard, L. (2003) Economic scenario. In: Foxell, S. (Ed.) *The professionals' choice: the future of the built environment professions.* London: CABE and RIBA. 55–80.

159. Curry and Howard (2003: 72).

160. Strelitz, Z. (2003) Social scenario. In: Foxell, S. (Ed.) *The professionals' choice: the future of the built environment professions.* London: CABE and RIBA. 99–118.

161. Strelitz (2003: 105).

162. Hughes, W. (2003: 87) Technological scenario. In: Foxell, S. (Ed.) *The professionals' choice: the future of the built environment professions.* London: CABE and RIBA. 81–98.

163. Hughes (2003: 98).

164. Nicolini, D. (2002: 167) In search of 'project chemistry'. *Construction Management and Economics,* **20**, 167–177.

165. Nicolini (2002: 175–176).

166. Pryke, S. and Smyth, H. (2006) *The management of complex projects: a relationship approach.* Oxford: Blackwell.

167. Ainamo, A. (2009: 224) Building working relationships with 'others'. *Building Research and Information*, **37**(2), 222–225.
168. Ibid.
169. Nicolini (2002), and; Gorse, C. A. and Emmitt, S. (2009) Informal interaction in construction progress meetings. *Construction Management and Economics*, **27**(10), 983–993.
170. Pietroforte (1997: 80–81).
171. For a more detailed critique, the reader is referred to Chan and Räisänen (2009)
172. See Greenwood *et al.* (2008).

Chapter 5

1. Personal communication with Dr. Martin Wood at the York Management School, University of York in 2008 about 'The fallacy of misplaced leadership'.
2. Wood, M. (2005: 1116–1118) The fallacy of misplaced leadership. *Journal of Management Studies*, **42**(6), 1101–1121.
3. We recognise the weakness in our methodology, in that we sought to speak to only the elite few of the UK construction industry. Furthermore, the nature of our participants is somewhat self-selecting in that they responded positively to our invitation to participate in our project. Virtually all of our participants have been educated up to University level themselves, and so this probably explains why there is greater alignment between their perspectives as practitioners and theoretical insights synthesised from the literature.

Index

Page numbers in *italics* represent figures, those in **bold** represent tables.

Accounting for People Task Force 81
Adamson, David 122
Advanced Institute of Management Research
    (AIM) 28
adversarial contracting 132
ageing population 18, 20
Amalgamated Society of Woodworkers 112
ambiguities 185–6
    see also paradoxes *and* tensions
American Society of Civil Engineers (ASCE) 10,
    13, 14–15
anchoring 5
arm's length management organisations 117
Asian Tiger economies 82
atmosfear 5
Australian Cooperative Research Centre for
    Construction Innovation (CRC) 10, 15–16
authority 175

Bebo 107
being alive 5
Bell, Sid 34
Big Ideas programme 49
biomimetics 15
Blair, Tony 144
Bloxham, Tom 35, 39, 52, 54, 65, 74, 114, 127,
    131–2
    biography 193–4
    tips for success 40
Blythe, Chris 32, 35, 36, 39, 68, 138, 140
    biography 190
bogus self-employment 60, 119–120
boundaries 180, 185

disruption of 184–5
Bradford 52–3
BrainReserve 4, 5
Broadgate project 33, 132, 189
Building Futures East 151–2
Building Schools for the Future 62
Bushnell, David 135
business relationships 13, 128, 130–3

CALIBRE model 80
Camden Market 35, 66, 194
capital
    human 15, 81–6
    man-made 78–81
    natural 86–93
    social 94–100, **96–7**
carbon emissions 92
cashing out 5
certificate 714 119
Chalmers University 10, 15
chaordic organisation 85
Chartered Institute of Building (CIB) 36
Chartered Institute of Personnel and
    Development (CIPD) 81
China 110
choice 29
citizen governance 149
    see also community governance
Civil Engineering Research Foundation
    (CERF) 10, 13–14
clanning 5
Clarke, Ken 116
climate change 20, 64–5, 86–7, **90**

cocooning 5
Coleridge, Samuel Taylor 31
Commission for Architecture and the Built
    Environment (CABE) 10, 12–13, 37, 38,
    158
communication 20
community engagement 155–73
community governance 147–50, **154**
    problems of 148–9
Conaty, Michael 151
Connect and Develop 107
Considerate Constructors Scheme 155
construction industry
    governance in 155–73
    interorganisational dynamics 162–5
    organisation of 161–2
    public image of 166–7
    structure 161–5
Construction Industry Council (CIC) 140
Construction Industry Institute (CII) 10, 13
Construction Industry Research and
    Information Association (CIRIA) 10,
    11–12
Construction Industry Training Board
    (CITB) 73, 114
Construction Research and Innovation Strategy
    Panel (CRISP) 9–11
ConstructionSkills 57
consultation 133–4
consumption 90–1
continuous improvement 74–5
corporate governance 145–7, **154**
corporate social responsibility 145–7
craft skills 135
critical events 35–7
critical trends/implications 16–17, 18–20
cross-national coordination 111–13

Department of Business, Innovation and Skills
    (BIS) 78
Department of Environment, Food and Rural
    Affairs (DEFRA) 121
Department of Trade and Industry (DTI) 121
Department of Work and Pensions (DWP) 121
depoliticisation of state 156–8
Dimbleby, David, *How We Built Britain* 94
diversity 71–2
    paradoxes of 113–15
down-ageing 5

Doyle, John 130
Doyle, Stephen 116
drivers for change 20

e-business 15
economic development 110–11
economic trends 11
education and training 70–5
    industry partnership 72–4
    rhetoric vs reality 82–3
    Scotland 72–3
    see also schools
embedded systems 15
emergent thinking 180, 183–4
employment structures 129–30
end-users, engaging with 133–4
energy 58–9
    ethics of consumption 90–2
Energy Saving Trust 38
Engineering and Physical Sciences Research
    Council 49
Enron 146
environmental agenda 87
environmental trends 11
Equal Opportunities Working Group 33
ergonomics 5
European Competition Directive 60
'eveolution' 5

Facebook 107
fantasy adventure 5
Fayol, Henri 28
Ferguson, George 32, 35, 36, 38, 39, 55, 58, 65,
    72, 114, 133, 136, 139
    biography 190
firm-specific skills 83
flash mobs 107
forecasting 4
foresight studies 7–21
    American Society of Civil Engineers
        (ASCE) 10, 13, 14–15
    Australian Cooperative Research Centre for
        Construction Innovation (CRC) 10, 15–16
    Chalmers University 10, 15
    Civil Engineering Research Foundation
        (CERF) 10, 13–14
    Commission for Architecture and the Built
        Environment (CABE) 10, 12–13
    Construction Industry Institute (CII) 10, 13

Construction Industry Research and
    Information Association (CIRIA) 10,
    11–12
Construction Research and Innovation
    Strategy Panel (CRISP) 9–11
critical appraisal 17, 21
critical trends and implications 16–17, 18–20
Frost, Robert 137
future agendas 7–9
future trends 6
futures thinking 3, 4, 6, 21, 179
    as emergent thinking 183–4
futuretense 5

generic skills 83–4
geography 34–5
Gilbreth, Frank 81
Gladwell, Malcolm, *The Tipping Point* 148
global economy 109
global-local tensions 55–8
globalisation 13, 16–17, 108, 142, 163, 180
    burden of coordination 111–13
    engaging with diversity 113–15
    economic and social development 110–11
Gore, Al, *An inconvenient truth: a global
    warning* 86
governance 67, 108–42
    citizen 149
    community 147–50, **154**
    in construction 155–73
    corporate 145–7, **154**
    depoliticisation of state 156–8
    joined-up 76, 120–2, 150–4, 182–3
    local 149
    measures of **154**
    political 143–5, **154**
    practitioner vs theoretical perspectives *175*
    shifting perspectives 142–54
    without government 143–5
government 59–65, 107–9, 115–122, 155–160
    clarity of intentions 64–5, 125
    continuity in 62
    devaluation of development 62–4
    as enabler 117
    evolving role of 108–9, 115–22
    as guardians of regulatory control 59–61
    provision of grassroots support 65–8
    regulation of private sector affairs 158–60
    relationship with construction 155–60

role in sustainable development 61–2
graduates in non-graduate occupations
    (GRINGOs) 82
Grand Designs 33, 38, 51, 191
green agenda 64

Handy, Charles 31
Hazlehurst, Guy 39, 40, 56, 59, 68, 111, 125,
    130, 134–5, 137, 138, 181
    biography 190–1
healthcare reform 152–3
Hemingway, Ernest 27
Hemingway, Geraldine 38
Hemingway, Wayne 38, 66–7, 114, 118, 134
    biography 194
Holdsworth, William 55
Homes for All 95
housing associations 117–18
Housing Corporation 118, 158
Huiden, Daniel 135
human behaviour, changing 92
human capital 15, 81–6
    complexity of skills definition 84–5
    generic vs specific skills 83–4
    rhetoric vs reality 82–3
    skills measurement 85–6
human relations 172–3
hybrids 184–5

icon toppling 5
individualism 20
industrial development 125
industry partnerships 72–4
industry response to sustainability agenda
    68–70
information technology 80
infrastructure 58–68
innovation 112
interactions 140–2
interdisciplinary thinking 135–6
interdisciplinary working 139, 185
intergenerational gap 70, 90–1
interorganisational dynamics 162–3
    inherent ambiguities 163–5
interprofessional working 185
intersectoral working 185
interview protocol 23
investing in people 81–6
Investors in People 36

joined-up governance 76, 120–2, 150–4, 182–3
  need for 150, 153

key reports 8
knowledge gaps 187
knowledge mobilisation 134–6
Kyoto protocol 92, 143
  see also climate change

Lamont, Norman 116
Lanarkshire 53
Landfill Tax 143
Lao Tzu 30
Latham, Michael 38–9, 60–1, 113, 115–18, 120,
    123, 124, 125, 127, 131, 132
  biography 192–3
Latham Report 33
leader-follower relationship 29
leadership 26–43
  as emergent process 181–2
  and innovation 30
  making of leaders 31–41
  practitioner vs theoretical perspectives *42*
  styles of 30
leadership theories 28–31
  application in construction 30
  critical antecedent people 31
  critical antecedent place 34
  critical antecedent events 35
  nature and nurture 37
learning 27
  passion for 40–1
legislation 158–60
Lenclos, Jean-Philippe 52
life cycle assessment 91
lifestyle expectations 20
Lipton, Stuart 132
Lloyds of London building 56
local governance 149
Lomborg, Bjorn 64
low-carbon agenda 64–5
Luebkeman, Chris 31, 34, 37, 40, 63, 65, 69, 71,
    109, 135–6, 137, 185
  biography 190
Luke, Ralph 41

McCloud, Kevin 33, 38, 39, 41, 51, 124, 133
  biography 191
MACE Limited 33, 36–7, 56, 189

McPherson, Ian 33
man-made capital 78–81
management 28
managerial functionalism 28, 30, 168–70
Manchester 54, 66
mavericks 38–9
Maynard, Bill 54
migrant workers 69, 160
migration 16–17
Monetary Policy Committee 49
Movement for Innovation 62
Mulgan, Geoff 144, 150
multi-skilling 134–5

nanotechnology 15
natural capital 86–93
  altering human behaviour 92
  environmental economics 87
  ethical dimension 90–2
  measurement of 89–90
  policy implementation 92–3
  strong vs weak sustainability 88
nature and nurture 37–41
networking 33
new entrants, failure to attract 71–2
Newborough, William 92
ninety-nine lives 5
non-governmental organisations (NGOs) 144

Obama, Barack 49, 64, 107
Olympic Development Authority 163
one-size-fits-all approach 114, 119
outsourcing 163
Owen, Robert 53

paradoxes 56, 76, 93, 106, 113, 174–5,185–6
  see also ambiguities *and* tensions
partnerships 130–3
  public-private *see* public-private partnerships
passion for learning 40–1
Pearce, David 48, 87
Pearce report 78
people
  influence of 31–2
  interactions 140–2
  senior 32–3
  working relationships 128–40
people and places 51–8
Phelps-Brown inquiry 162

place, importance of 34–5
pleasure revenge 5
political governance 143–5, **154**
political leadership 127
political trends 11, 20
Popcorn, Faith 4–5
power, shifting of 186
private finance initiatives 16, 105, 116, 125, 157
private sector
    government regulation of affairs 158–60
    increasing involvement 118–20
    rising role of 115–22
    state engagement with 156–8
privatisation 155–73
productivity
    measurement of 80–1
    residual 81
productivity comparisons
    difficulties of 79
    limitations for economic policy 79–80
professionalism 136–40, 165–71
    engagement in society 167–8
    fragmentation of 167–8
    future of 170–1
    managerial agenda 168–70
    reinforcing boundaries 166–7
profit 123–5
project chemistry 172
Public Interest Disclosure Act (1998) 170–1
Public Sector Borrowing Requirement 157
public-private partnerships 105, 115–18, 122–8, 157
    aims of 125–8
    long-term view 125
    social responsibility vs profit 123–5
Putnam, Robert, *Bowling alone* 107

Raynsford, Nick 34, 37, 39, 61, 112, 117, 121–2, 140
    biography 192
rebels 38–9
registered social landlords 117, 118, 158
representation 175
research 70–5
residual productivity 81
Respect for People 155
reward-punishment framework 29
Rhys Jones, Sandi 32, 33, 36, 38, 39, 71, 74–5, 135, 136, 137

    biography 192
Rippon, Geoffrey 116
Ritchie, Alan 32, 34, 35, 53, 60, 70–1, 112, 115, 119, 128
    biography 189
Rogers, Peter 132
Rogers, Richard 56, 95
Rouse, Jon 34, 37, 38, 40, 52, 72, 109, 111, 118, 121, 122–3, 133, 138
    biography 53–4, 191
Royal Institute of British Architects (RIBA) 170
Ryrie Rules 116

Salt, Titus 53, 94
Saltaire 52, 53, 94
Save Our Society (SOS) 5
schools 126
    private finance initiatives 127
Scotland, education 72–3
Scottish building Apprenticeship Training Council 73
Scottish Parliament building 56, 60
Second Health 152–3
Senge, Peter 29
senior people, access to 32–3
Shelton, David 41
short termism 68, 71
Simms, Neville 116
skills
    complexity of definition 84–5
    firm-specific 83
    generic vs specific 83–4
    measuring 85–6
skills agenda 82–3
skills mobilisation 134–6
small indulgences 5
small- to medium-sized enterprises 73, 141
smart materials 15
Smith, Adam 81–2
Snow, Charles Percy 162
social capital 94–100, **96–7**
    community governance 147–50
    top-down engagement 98–100
    see also sustainable communities
social development 110–11
social housing 124
social network theory 164–5
social networking websites 107
social responsibility 123–5

Social, Technological Economic,
    Environmental and Political (STEEP)
    framework 11
social trends 11
Solow, Robert 81
Spain, labour market regulation 159
specific skills 83–4
Stefanou, Stef 32, 35, 41, 62, 64, 69, 75, 119, 121,
    126, 129–30, 136–7
  biography 193
Stern report 87
strong sustainability 88
  elusiveness of 88–9
Suede, Henry 34
sustainability 50
  people and places 51–8
  strong vs weak 88
sustainable communities 51–2, 94–100, **96–7**
  community participation 95, 98
  definitions of 95, **96–7**
  economic vs social aspects 52–3
  elements of 52
  interactions between people 54–5
  success of 55–8
sustainable consumption 90
sustainable development 47–104, 182
  definitions 78–102
  education and research 70–5
  industry response 68–70
  infrastructure 58–68
  measurement problem 100–2, *101*
  perspectives *77*, 78–102, *101*, *103*
  role of political leaders 58–68
sustainable energy sources 58–9
sustainable skills landscape 56

Taylor, David 41

technological trends 11, 20
technology 13
temporary project coalition 163
tensions 55, 69–71, 76, 93, 174–6, 185–6
  see also ambiguities *and* paradoxes
terrorist threats 20
Thatcher, Margaret 61
*The professionals' choice: the future of the built
    environment professions* 170
theoretical models 80–1
think global act local 109–15
top-down engagement 98–100
trade unions 129

vigilante consumers 5

weak sustainability 88
White, Bob 33, 34, 36–7, 56, 62, 63, 73, 74, 110,
    111–12, 126, 130, 132, 134, 138–9
  biography 189
women in construction 71–2
Woods-Waters, Anthony 151
work-time model 80
workforce 13
working relationships 128–40
  engaging with end-users 133–4
  mobilising skills and knowledge
    134–6
  partnerships 130–3
  professionalism 136–40

Yeang, Kenneth 38, 69, 124
  biography 191
  professional architects 139
Youtube 107

zero-carbon housing 92